Don't Tear it Down!

Preserving the Earthquake Resistant Vernacular Architecture of Kashmir

Text and Photographs by Randolph Langenbach

United Nations
Educational, Scientific and
Cultural Organization

Produced by UNESCO New Delhi Office
B-5/29, Safdarjung Enclave
New Delhi - 110 029
India
+91 11 2671 3000
newdelhi@unesco.org
www.unesco.org/newdelhi

Minja Yang, Director, UNESCO New Delhi Office
Nicole Bolomey, Programme Specialist for Culture
Ahmed Fahmi, Programme Specialist for Natural Sciences
In association with:
United Nations Development Programme (UNDP)
Government of Jammu and Kashmir
Kashmir Earthquake Relief
UN-HABITAT

Prepared for UNESCO by:

Randolph Langenbach
Conservationtech Consulting
6446 Harwood Avenue
Oakland, California 94618
USA
+1-510-428-2252
www.conservationtech.com

With contributions from:
Hakim Sameer of INTACH J&K,
Tom Schacher of SDC, and Maggie Stephenson of UN-HABITAT.

Referenced website, **www.traditional-is-modern.net**

The recommendations herein are intended to be used for general educational purposes, and are not meant to be a substitute for competent professional analysis and advice for the undertaking of individual projects. The authors and UNESCO accept no liability for the structural behaviour of any structure or for work undertaken that may be based on the contents of this report.

The designations employed and the presentation of material in this information product and on maps do not imply the expression of any opinion whatsoever on the part of the UNESCO concerning the legal or development status of any country, territory, city or area or of its authorities, or concerning the delimitation of its frontiers or boundaries.

Book Design: Susanne Weihl: folio2, Palo Alto, California, USA
Prepress Production: JoAnn Kolonick, San Francisco, California.
Printing: Silverline Communications, New Delhi

Publisher's Cataloging-In-Publication Data
(Prepared by The Donohue Group, Inc.)

Langenbach, Randolph, 1945-   Don't tear it down! : preserving the earthquake resistant vernacular architecture of Kashmir / text and photographs by Randolph Langenbach. -- 1st American ed.

    p. : ill., charts, maps ; cm.

    Originally published in India by United Nations Educational Scientific and Cultural Organization (UNESCO).
    Includes bibliographical references and index.
    ISBN-13: 978-0-9796807-1-7
    ISBN-10: 0-9796807-9-9

1. Earthquake resistant design--India--Jammu and Kashmir. 2. Buildings--Earthquake effects--India--Jammu and Kashmir. 3. Earthquakes--India--Jammu and Kashmir. 4. Jammu and Kashmir (India)--Buildings, structures, etc. I. Title.

TA658.44 .L36 2009
624.1/762/0954

Printed in India

At this stage I made inquiries about the country and city of Kashmír from men who were acquainted with it, and from them I learned that Kashmír is an incomparable country. In the midst of that country there is a very large and populous city called Naghaz. The rulers of the country dwell there.

The buildings of the city are very large and are all of wood, and they are four or five storeys high. They are very strong and will stand for 500 or 700 years.

A large river runs through the middle of this city, as large as the Tigris at Baghdád, and the city is built upon both sides of it. The source of this river is within the limits of Kashmír in a large lake,...which is called Vír-nák. The inhabitants have cast bridges over the river in nearly thirty places. These are constructed of wood, stone, or boats; seven of the largest are within the city, and the rest in the environs.

Tímúr the Tátár, 1398
(Elliot 1867)

1. Narrow lane between old shops and houses on the left bank of the River Jhelum north of 5th Bridge, Srinagar, 2005.

Contents

Taq: Timber-laced masonry bearing wall construction

Taq construction is a composite system of building construction with a modular layout of load-bearing masonry piers and window bays tied together with ladder-like constructions of horizontal timbers embedded in the masonry walls at each floor level and window lintel level. These horizontal timbers tie the masonry in the walls together, thus confining the brick mud or rubble stone of the wall by resisting the propagation of cracks. The masonry piers are almost always 1 to 2 feet square and the window bay/alcove *(taqshe)* 3 to 4 feet in width. The *taq* modular layout defines the Kashmiri house size measurements, i.e. a house can be 3 *taq* (window bays) to 13 *taq* in width. In Pakistan, timber-laced masonry is known by the Pashto word *bhatar*.

Dhajji Dewari: Timber frame with infill masonry construction

Dhajji dewari is a timber frame into which one layer of masonry is tightly packed to form a wall, resulting in a continuous wall membrane of wood and masonry. The term is derived from a Persian word meaning "patchwork quilt wall". The frame of each wall consists not only of vertical studs, but also often of cross-members that subdivide the masonry infill into smaller panels, impart strength and prevent the masonry from collapsing out of the frame.

There are a number of other types of wood frame and wood wall vernacular construction types that can be found in Kashmir on both sides of the Line of Control, but these are not covered in this volume. For more information on these and other subjects related to this book, please see www.traditional-is-modern.net.

FOREWORD

UNESCO's Kashmir initiative was begun by UNESCO New Delhi in cooperation with UNESCO Islamabad following the tragic earthquake of October 8, 2005. It was undertaken in collaboration with The Indian National Trust for Art and Cultural Heritage – Jammu & Kashmir Chapter (INTACH J&K) and students from the Bangalore School of Architecture. Following an initial reconnaissance of damage to heritage structures, UNESCO identified a need for major advocacy to support the preservation of Kashmiri vernacular buildings, not only for their heritage value, but also to disseminate knowledge of their uniquely valuable construction techniques and resilience against earthquakes.

International experts on earthquake engineering and cultural heritage from around the world gathered in March 2006 to discuss this project and to review a draft repair manual commissioned by UNESCO and prepared by the National Centre for Peoples'-Action in Disaster Preparedness (NCPDP), Ahmedabad, and this was followed by a UNESCO-organized workshop in Srinagar in June 2007. Also in 2007, UNESCO commissioned Randolph Langenbach of Conservationtech Consulting in the United States to prepare this publication. Prof. Langenbach, who has worked on Kashmir and traditional construction in earthquake areas for over 25 years, was asked to bring greater focus to the heritage values of Kashmiri architecture and traditional construction techniques and knowledge systems. *Don't Tear It Down! Preserving the Earthquake Resistant Vernacular Architecture of Kashmir* is the product of this work.

This publication focuses on the historical construction systems of these vernacular buildings in an effort to offset the common belief that these systems are obsolete and inadequate for modern-day life. It is meant to encourage conservation of Kashmir's vernacular architecture through increased understanding of its scientific and cultural attributes, as well as its earthquake resistant features. It is our hope that the information and guidelines given in this publication will be circulated widely to raise the awareness of architects, engineers and citizens at large, so that present and future generations will benefit from the knowledge of the past to make informed choices that will create a safer, as well as culturally rich, built environment.

This volume, however, is offered as more than just a technical document. It also aims to celebrate the traditional architecture, urban landscape, and life of Srinagar. It is intended to convey a vision of the cultural value of Srinagar and of Kashmir in general, and thus to encourage the people of the city and the region to preserve the cultural heritage of this remarkable place for themselves – and for the world.

Minja Yang, Director
UNESCO New Delhi Office
February 2009

Acknowledgements

This publication has been printed thanks to the generous participation of the United Nations Development Program (UNDP), the United Nations Human Settlements Programme (UN-HABITAT) and Kashmir Earthquake Relief (KER) (www.kashmirrelief.org). Additional support has been generously provided by Funkar International Inc. (www.funkar.org), Global Urban Development (www.globalurban.org), Rajinder Chaku, Obaid Z. Khan, Shobi S. Khan, Tasawar Jalali, and other donors.

UNESCO and the author would like to thank the Ministry of Science and Technology, Ministry of Home Affairs, Government of India, the State Government of Jammu & Kashmir, the UNDP, UN-HABITAT, and INTACH J&K for their assistance, financial and in other ways, and their kind cooperation in developing this publication and in conducting the related workshop in Srinagar in June 2007. Thanks go also to the Archaeological Survey of India, Government of India, for their generous facilitation of Prof. Langenbach's reconnaissance mission to the damage district in Kashmir after the October 2005 Kashmir earthquake, and to the Earthquake Engineering Research Institute (EERI) and the US National Science Foundation for their support of the reconnaissance trip to Kashmir, as well as earlier research in Turkey, through their Learning from Earthquakes Program.

The author's work on Kashmir and on traditional construction in earthquake areas in general also has been supported by the University of California Berkeley's Committee on Research, the American Academy in Rome, the International Centre for the Study of the Preservation and Restoration of Cultural Property (ICCROM), and the Samuel H. Kress Foundation through the US National Committee of the International Council on Monuments and Sites (US/ICOMOS).

We wish to also thank the experts from India and abroad who have helped this project through all of its stages, from the expert meeting in March 2006 until the last edits in March 2009, for their dedication in working with UNESCO and the author in this activity. This includes Saleem Beg, Hakim Sameer Hamdani, and Jabeen Manzoor of INTACH J&K; Professor G.M. Bhat, Department of Geology, Jammu University, Jammu; Professor Kaiser Bukhari, National Institute of Technology, Srinagar; Iftikhar Jalali, Educator, Srinagar; Tasawar Jalali, Executive, USA; Dr Rohit Jigyasu, Consultant to UNESCO; Rafique A. Khan, Urban Planner, USA; Tasneem F. Khan, Kashmir Earthquake Relief, USA; Rupal and Rajendra Desai of the National Centre for Peoples'-Action in Disaster Preparedness (NCPDP), Ahmedabad, and the participating students of Bangalore School of Architecture; Tom Schacher of the Swiss Agency for Development and Cooperation (SDC) in Pakistan; Maggie Stephenson of UN-HABITAT in Pakistan; and many others who participated in this initiative.

In addition to those above, the author would like to express his personal thanks to the following friends and colleagues who have given generously of their time and expertise to provide advice and comment on the manuscript for this work: Professor Anand Arya, Engineer, New Delhi; Professor Hassan Astaneh, Engineer, USA; James Baer, Copy Editor, UK; Jitendra Bothara, Engineer, Nepal and New Zealand; Rajinder Chaku, Architect, Kashmir and Canada; Professor Andrew Charleson, Engineer, New Zealand; Vítor Cóias e Silva, Engineer, Portugal; Craig Comartin, Past President, EERI, USA; Tom Dolan, Architect, USA; Alberto Dusi, Engineer, Italy; Marjorie Greene, EERI, USA; Professor Polat Gülkan, Engineer, Turkey; Professor Robert Hanson, USA; Frederick Hertz, Attorney, USA; Kubilây Hiçyılmaz, Engineer, UK; Richard Hughes, Conservator, UK; John Hurd, Conservator, UK; Professor Bilge Işik, Engineer, Turkey; Professor Sudhir Jain, Engineer, India; Anup Karanth, Disaster Recovery Specialist, India; Akshay Kaul, Landscape Architect, Kashmir and New Delhi; Professor Alessandra Marini, Engineer, Italy; Professor C.V.R. Murty, Engineer, India; Carlos Sousa Oliveira, Engineer, Portugal; Marla Petal, Disaster Relief Specialist, USA; Professor Durgesh Rai, Engineer, India; Hemen Sanghvi, Architect, Gujarat; Tom Schacher, Architect, Switzerland; Amir Soltani, Journalist, Iran and USA; Maggie Stephenson, Architect, Ireland; Niranjan Swarup, Engineer and Director, New Delhi; Stephen Waite, Architect, Seattle; Professor Tim Winter, Sociologist, Australia; Marc Weiss, Urban Planner, USA; and Peter Yanev, Engineer, USA.

2. Persian influence can be seen here in the Jalali Haveli, Srinagar, of taq construction. It was constructed by a family of Persian descent. A date inscribed on one of the wooden beams is 1280 in Hijri, an Islamic calendar, which was being used in Kashmir until the last century. This date translates to 1863 in the Gregorian calendar.

3. 1979 view of the River Jhelum showing the plethora of doongas lining the banks beneath the tightly clustered gable-roofed timber and masonry dwellings, Srinagar, Kashmir. Photo by Tom Dolan.

4. Srinagar, 2005. Srinagar is one of the world's last capital cities where the towers of the religious structures are the tallest buildings in the city. It is a remarkable asset that needs to be consciously protected if it is to remain intact.

5. Dhajji dewari *construction in central Srinagar.*

INTRODUCTION

Earthquakes have occurred regularly over centuries in Kashmir and people have learnt to live with it. Two old construction systems known as taq *and* dhajji-dewari *exist here side-by-side and both have tested quake-resistant features.*

Centre for Science and Environment (CSE)
Gobar Times, 15 February 2006[1]

The Himalayan mountain chain that surrounds Kashmir separates the Indian subcontinent from the Tibetan Plateau. It extends across six nations: Afghanistan, Bhutan, China, India, Nepal and Pakistan. In geological terms it is a young mountain chain. It continues to be uplifted by the movement of the earth's tectonic plates – a movement that has given it the world's highest peaks, and which continues to result in periodic earthquakes across a seismically active belt that includes Kashmir. The 2005 Kashmir earthquake has served to renew people's awareness of this risk.

The Vale of Kashmir is located in the western part of the Himalayan mountain range on the site of a prehistoric lake created by the uplift of the mountains between Indian and Pakistan Administered Kashmir. Over geological time, this lake gradually silted in, and the alluvium from the mountains became the fertile soil of the valley floor. This is responsible both for the area's rich agriculture and for its earthquake vulnerability. Srinagar lies on one of the most waterlogged soft soil sites for a capital city in the world.

The Indian and Pakistan administered areas of Jammu and Kashmir.

The recorded cultural history of Kashmir dates back 3,000 years. The oldest known remains of monumental buildings are the earthquake-damaged ruins of early Hindu and Buddhist temples built of large blocks of stone. Later medieval structures, some of them religious buildings constructed by the Muslim community, were made of a more economical and lightweight combination of mud, stone and brick, well tied together with timber. This construction system with its use of masonry laced together with timber, which is mentioned in texts from the 12th century, was the beginning of the urban architecture in the Vale of Kashmir as we know it today. In our time, Srinagar and other cities and villages in Kashmir are distinguished not only by their great monuments, but first and foremost by their vernacular residential architecture. It is an architecture generated out of a distinctive use of materials and way of building which are adapted to the local climate, culture and natural environment, principally the soft soils and the earthquake risk in the region.

The danger of earthquakes and the soft building ground have had a great influence on the way people traditionally built their houses. This combination of soft soils with earthquakes requires buildings that can undergo a certain amount of inelastic deformation without losing their vertical load-carrying capacity. Rigidity in a construction carries the potential for destruction. The more rigid a building is, the stronger (and thus more expensive) it must be

6. Large taq *house on Rainiwari Canal, Srinagar, 2005.*

7. Stairway in Bhat Family House, Pampore (see also Figures 5.1e & 5.1f).

8. Interior of Jalali Haveli, Srinagar (see also Figure 2).

in order to avoid damage as the ground moves or settles. Historically, because of the available traditional materials and means of construction in Kashmir, strength was not always possible, so a certain amount of flexibility or "give" was essential.

There are so many influences on the development of building construction traditions that it is not easy to isolate any one reason for the timber lacing in the masonry, but its effectiveness in holding the masonry together on soft soils undoubtedly has played a major role. It has also proven to be effective in reducing damage in earthquakes, which may help explain why variations of it can be found in the mountains, where soft soils are not a problem. At the beginning of the 19th century the systems evolved to become the two main traditional construction systems: *taq* (timber-laced masonry) and *dhajji dewari* (timber frame with masonry infill). In Pakistan, timber-laced masonry is known by the Pashto word *bhatar.*

This remarkable vernacular architectural heritage of Kashmir is, however, under threat and is rapidly being lost, especially now, following the earthquake of 2005. Its inherent qualities and great architectural expression, together with its unique construction, are insufficiently recognized or considered important by the citizenry today. Thus this architecture is rapidly being displaced by non-indigenous reinforced concrete buildings, many of which are constructed in a way that has proven to be particularly dangerous in earthquakes, as was demonstrated in 2005.

In addition, the cement plants which have sprung up to supply this insatiable demand have contributed to an assault on the air quality and the environment that threatens to irretrievably diminish the beauty for which Kashmir has been famous for centuries. Too few people today recognize how much Kashmir's traditional residential buildings create the unique character of urban Kashmir and complement the magnificent natural landscape with an equally rich cultural tradition.

In the current age, when energy conservation and the effects of greenhouse gasses have come to the forefront in international debates over humankind's future on the planet, it is important to understand that the conservation of historic buildings can play a particularly important role in any environmental conservation effort. This is discussed in the last chapter. Many have said that timber is too expensive, even when it comes to the maintenance of existing buildings with timber in their construction. Yet wood is the most renewable and least energy consumptive resource that one can use for building construction. In addition, when the timber already in a heritage structure is left in place and protected from decay, it helps to avoid the loss of timber that is still on the stump. The embodied energy and material assets in existing buildings are considerable, as represented by the energy and resources consumed for their replacement.

More important, however, is the cultural and economic loss that comes from the destruction of heritage buildings. If one looks purely at the economics, the benefits are almost always apparent. Citing examples of city centre revitalization efforts in the United States when speaking to Europa Nostra, the pan-European Federation for Cultural Heritage, in 2005, international development specialist Donovan Rypkema observed: "I cannot identify a single example of sustained success in downtown revitalization where heritage conservation wasn't a key component. Conversely, the examples of very expensive failures in downtown revitalization have all had the destruction of historic buildings as a major element." After explaining how extraordinarily important heritage conservation is to the economic well-being of communities, he went on to say: "In the long run the economic impact of heritage conservation is far less important than its educational, environmental, cultural, aesthetic, and social impact" (Rypkema, 2005). Preservation of the vernacular architecture of Srinagar and other cities and towns in Kashmir is of critical importance for all of these reasons. Because the architectural expression and the construction techniques of traditional Kashmiri buildings are so closely intertwined, conservation cannot succeed without first addressing the structure and construction systems themselves.

Chapter 1 of this book identifies the traditional construction that is found in Srinagar and the other urban areas of Kashmir, focusing on the *taq* and *dhajji dewari* construction with a brief description of its earthquake resistant attributes. Chapter 2 provides a general primer on earthquake engineering concepts, as these will help in understanding specifics about the safety and performance of buildings in earthquakes, and how they can be repaired and strengthened. Knowledge of these concepts will help readers understand how traditional *taq* and *dhajji dewari* construction actually work, why they work as well as they do, and thus why the heritage buildings in Kashmir should and can be preserved. (A glossary of scientific and engineering terms to help with this and the other chapters is included as an appendix.)

Chapter 3 delves into the earthquake resistant features of *taq* and *dhajji dewari* construction in detail with comparison to similar construction in Gujarat, India, in Pakistan (including the Northern Areas, and Pakistan Administered Kashmir), and in other parts of the world. Chapter 4 provides specific advice on the repair and strengthening of *taq* and *dhajji dewari*, together with a procedure for analysing specific buildings for conservation and seismic resistance. Chapter 5 concludes with a discussion of present construction and development trends in Kashmir and Srinagar in particular, and makes the case for the conservation of traditional structures and the revival of traditional building techniques.

9. 1864 view by Samuel Bourne of the River Jhelum, Srinagar, from Fateh Kadel (2nd Bridge), showing the Shah Hamadan Mosque (Khanquah-e-Moualla) and houses.

10. River Jhelum from Fateh Kadel (2nd Bridge) taken in 2005 close to where the 1864 image was taken. Many of the houses are still of traditional construction.

1a Houses of traditional construction on the River Jhelum next to Old Habba Kadel (5th Bridge) in Srinagar.

KASHMIR'S EARTHQUAKE RESISTANT
TRADITIONAL CONSTRUCTION

In the town are many lofty buildings constructed of fresh cut pine. Most of these are at least five storeys high....[causing viewers to] bite the fingers of astonishment with the teeth of admiration...

Mirza Haider Dughlat
Tarikh-i-Rashidi, 16th century

As was briefly mentioned in the Introduction, most of the traditional buildings in Srinagar and the Vale of Kashmir can be divided into two basic systems of construction. The first system, *taq*, consists of load-bearing masonry walls with horizontal timbers embedded in them. These timbers are tied together like horizontal ladders that are laid into the walls at each floor level and at the window lintel level. They serve to hold the masonry walls together and tie them to the floors. The second system, *dhajji dewari* construction, consists of a braced timber frame with masonry infill, which for example in England is commonly referred to as "half-timbered" construction.

Not much is known about when *taq* or *dhajji dewari* came into vogue, but the notion of using timber members as braces within weak masonry walls can be found in many cultures through history. In Kashmir, historical sources can be found which remark on the timber construction from as early as 1148 by Kalhana in his *Rajtarangani* ("Chronicle of Kings") who said the "mansions of the city...reached the clouds and were mostly built of wood," and again in 1398 by Tímúr the Tátár, who wrote in his autobiography that "the buildings of the city are very large and all of wood" and they are "very strong and will stand for 500 or 700 years" (Elliot, 1867). In the 16th century, Muhammad Haidar Dughlát in his *Tarikh-i-Rashidi* ("A History of the Moghuls in Central Asia") remarked, "In the town are many lofty buildings constructed of fresh cut pine. Most of these are at least five storeys high.... [causing viewers to] bite the fingers of astonishment with the teeth of admiration" (Bamzai, 1994).

While these accounts only comment on the use of wood, rather than masonry integrated with wood or confined in a wooden frame, before the advent of modern saws and nails some form of masonry most likely would have been used to enclose the structures. Thus, these accounts may be describing the ancient precursors of either *taq* or *dhajji dewari* construction. Different variations of *taq*, with its masonry walls with horizontal timber lacing, may have been in common usage throughout the Himalayan region before *dhajji dewari* came into general use, but *dhajji* construction, which was embraced by the British because of its resemblance to the Elizabethan half-timbered construction common in Britain, has continued to be used more often than *taq* in recent years.

In some remote mountain areas in Pakistan's Northwest Frontier Province to the west and north of Kashmir, a version of *taq*, there called *bhatar* (plural: *bhatari*), continues to be used for new construction. The Pashto word *bhatar* more precisely describes the construction found in *taq* buildings because it specifically refers to the horizontal beams in the walls. These usually have a cross-section of 4"x4" (10 cm square) to 6"x6" (15 cm square). While *bhatar* shares the same method of timber reinforcing as *taq*, unlike *taq*, the masonry is not configured into a system of thick piers and thinner unbonded panels (see Section 1.1) The masonry in *bhatar* is usually stone, and often small broken

1b Exterior detail of cribbage construction of 18th-century Shah Hamadan Mosque (Khanqah-e-Moualla), Srinagar.

1c Example of cator *and cribbage construction in the Hunza Fort in Karimabad, Pakistan after restoration by the Aga Khan Cultural Services Pakistan. Photo by Tom Schacher, 2007.*

stones are laid into the walls without mortar, which makes the timber reinforcements all the more important. Examples of this kind of construction (usually with a clay mortar) can also be found in the mountain areas of Kashmir, on both sides of the Line of Control, as well as the northern Indian states of Himachal Pradesh and Uttarakhand, near to Kashmir.

Several of the historic mosques in Srinagar are of "cribbage" construction, a variation of timber-laced masonry construction that can be found in the Himalayan mountains of northern India, northern Pakistan near the Chinese border, and parts of Afghanistan (see Figures 1b and 1d). This has proven to be particularly robust in earthquake-prone regions, but as wood supplies became depleted it must have been found to be extravagant. This may in part explain the origins of the *taq* and *bhatar* systems, where the timber lacing is limited to a series of horizontal interlocking timber bands around the building, thus requiring significantly less wood in its construction.

A combination of cribbage at the corners with timber bands, known as "*cator* and cribbage", can be found in the Hunza region of Northern Areas of Pakistan (Figure 1c). Examples can also be found in the Himalayan regions of northern India. This is a heavier, more timber-intensive version of timber-laced masonry than *taq* that dates back some 1,000 years (Hughes, 2000). The corners consist of a cribbage of timber filled with masonry. These are connected with timber belts (*cators*) that extend across the walls just as they do in *taq* and *bhatar* construction.

There is evidence that many of these construction traditions have followed patterns of migration and cultural influence over centuries, such as the spread of Islamic culture from the Middle East across Central Asia, including Kashmir and other parts of India. In Turkey, timber ring beams in masonry, known singly as *hatıl* and plural *hatıllar*, are part of a construction tradition that is believed to date back 9,000 years (Hughes, 2000). The Turkish word *hatıl* has the same meaning as *cator* does in Balti language. Also in Turkey, another common traditional construction type, *hımış,* is essentially the same as *dhajji* construction in Kashmir.

British conservator Richard Hughes has noted that "The use of timber lacing is perhaps first described by Emperor Julius Caesar as a technique used by the Celts in the walls of their fortifications. Examples, with a lot of variations, are to be noted from archaeological excavations of Bronze and Iron Age hill forts throughout Europe."

Hughes also cites examples in the Middle East, North Africa and Central Asia (Hughes, 2000). Different variations on all of these construction types are also likely to be found in the areas outside of the regions discussed in this volume, including Nepal, Bhutan and parts of China, including Tibet.

1d *The Khankah in Pampore, near Srinagar, was constructed ca.1600. The 3 foot (1 metre) thick walls are of heavy timber cribbage construction. This unusual and impressive interior space is of world importance. (At the time this photograph was taken in May 2007, plans were underway to remodel this room by covering all of the historic surfaces with plywood.)*

1e *View of the anteroom shown above in the Khankah in Pampore, where modern plywood was being installed over the ancient heavy timber cribbage walls in May 2007, in an effort to "modernize" the interior.*

1.1a *18th- or early 19th-century canal shop house alongside what was once the Mar Canal in Srinagar. This building survived the 1885 earthquake. The timber lacing has held this structure together despite extensive differential settlement. Note that the windows are not lined up vertically: this feature would not be structurally possible had the walls not been reinforced with timbers.*

1.1b Taq *building on the bank of the River Jhelum. The fact that this bearing wall masonry structure can be held up on three poles is a testament to the effectiveness of the timber lacing. The shoring poles are supporting the structure under the three main beams that support the joists, two of which are embedded in the front and back masonry walls.*

1.1. *TAQ* CONSTRUCTION

A combination of wood and unreinforced masonry laid on weak mortar gave [taq] buildings the required flexibility. The wooden bands tied the mud mortar walls and imparted ductility to an otherwise brittle structure. Built by masons, who had no formal degrees in structural engineering and architectural design, these structures stand today as the epitome of human creative instincts. Yet, these buildings continue to fascinate modern-day engineers... Tragically, however, rarely does this learning translate into constructions based on such masterly designs.

Sr. Sudhirendar Sharma, Development Analyst & Ashoka Fellow
Ashoka Changemakers, 2005

Taq construction is a bearing wall masonry construction with horizontal timber lacing embedded into the masonry to keep it from spreading and cracking. It is usually configured with a modular layout of masonry piers and window bays tied together with ladder-like constructions of horizontal timbers embedded in the masonry walls at each floor level and window lintel level. The masonry piers are thick enough to carry the vertical loads, and the bays may either contain a window, or a thinner masonry wall as required by the floor plan and the building's orientation. The ladder-like sets of timber beams *(ker)* laid into the exterior and interior faces of the walls are connected together through the wall either by the floor beams *(veeram)* and joists or short connector pieces (see Figure 1.1h for a view of similar construction in Ahmedabad where the connector pieces are visible). These horizontal "ladder bands" are located at the base of the structure above the foundation *(das* or *dassa)*, and at each floor level and at the window lintel level.

The face bricks traditionally used during the 19th and early 20th centuries were small in size, rough-surfaced, and hard-fired (Figure 1.1f). They are known as Maharaji bricks because the Maharaja of the Dogra period (1846–1947) monopolized their production. They served as the weather-resistant skin over sun-dried brick (known locally as *khaam seer*) or rubble made up of broken bricks laid in clay mortar (Hamdani, 2006).

There is no specific name in Kashmiri to identify this timber-laced construction method itself, but the closest name used to describe it is *taq* because this is a name for the type of buildings in which it is commonly found. *Taq* refers to the modular layout of the piers and window bays, i.e. a five-*taq* house is five bays wide. The masonry piers *(tshun)* are almost always 1½-2 feet (45-60 cm) square, and the bays are approximately 3-4 feet (90-120 cm) in width. Because this modular pier and bay design and the timber-laced load-bearing masonry pier and wall system go together, the name has come to identify the structural system as well.

1.1c Taq *construction in Srinagar. The symmetrical layout of windows is characteristic of* taq *and it is from this that the name of the construction system is derived.*

An important factor in the structural integrity of *taq* is that the full weight of the masonry is allowed to bear on the timbers, thus holding them in place, while the timbers in turn keep the masonry from spreading. The spreading forces can result over time from differential settlement – or in an instant in an earthquake. The overburden weight of the masonry in which the timbers are embedded serves to "pre-stress" the wall, contributing to its resistance to lateral forces.

The best early account of the earthquake performance of *taq* construction maybe the one by British traveller Arthur Neve, who was present in Srinagar during the earthquake of 1885 and published his observations in 1913:

> *"The city of Srinagar looks tumbledown and dilapidated to a degree; very many of the houses are out of the perpendicular, and others, semi-ruinous, but the general construction in the city of Srinagar is suitable for an earthquake country; wood is freely used, and well jointed; clay is employed instead of mortar, and gives a*

1.1d House under demolition showing how the masonry wall of taq is subdivided into panels where the light shows through between the wall panel and the half-demolished pier.

1.1e Photograph taken in the 1980s showing a building under demolition with taq timbers visible at the window lintel level and at the floor level. The floor joist can be seen to penetrate the wall where it is sandwiched between the exterior runner beams.

somewhat elastic bonding to the bricks, which are often arranged in thick square pillars, with thinner filling in. If well built in this style the whole house, even if three or four storeys high, sways together, whereas more heavy rigid buildings would split and fall." (Neve, 1913)

One most unusual element in the Kashmiri *taq* system is the existence of a deliberately unbonded butt joint between the masonry piers and the wall and window panels that can be seen in Figure 1.1d – a seemingly irrational detail which one would think would weaken the building wall in an earthquake. Nevertheless, it is a common, yet distinctive, feature in many (though not all) *taq* buildings. This joint divides the masonry walls into piers that are physically separate from the layer of masonry surrounding the windows or spanning the bay where there are no windows.

This system, with the timber tie beams together with the masonry subdivided into piers and panels, was most likely invented to avoid diagonal tension cracks from differential settlement of the foundations on the soft soils of the former lake bed that would otherwise have

1.1f View of the mid-19th century Jalali Haveli in Srinagar showing Maharaji bricks. Maharaji bricks, common in the 18th to early 20th centuries, were shaped without the use of a wooden mold and thus are of varying length and about 1 inch (2.5 cm) in thickness.

disrupted the masonry, as can be seen in Figure 1.1g. It may also have evolved as a construction sequencing method, allowing the early completion of the roof, which rests on the piers, before the rest of the walls and interior was completed. In barns and the top storey on rural houses, the space between the piers is left open when the floor is used as a hay loft.

In earthquakes, the value of these joints in the masonry is more questionable. From a positive standpoint, the separation of the panels of masonry serves as a crack stopper, and the avoidance of settlement cracks reduces the likelihood that the wall would be weakened by them before an earthquake, but subdivision of the wall into panels would seem to make it less resistant in an earthquake than if the masonry panels were bonded to the piers. In fact when one examines the actual construction of the masonry common in the older houses, with its amorphous clay- and rubble-filled cores (Figure 1.1j), the crack-stopping control joints are probably effective in reducing the negative effects that these cores would otherwise have if they were not limited to discrete elements. The timber ladder bands and timber floors bridge across the joints, keeping the individual piers from separating and the house from breaking apart.

Taq construction has largely gone out of use in Indian Administered Kashmir for new construction, but in Pakistan's Northwest Frontier Province, however, new buildings with *bhatar* and *cator* and cribbage construction continue to be constructed, particularly since the 2005 earthquake focused renewed interest on the seismic resistance of these systems – especially in preference to plain rubble stone without timber lacing, which had been commonly used in rural areas until the earthquake. In addition, *bhatar* has been successfully reintroduced into remote areas of Pakistan Administered Kashmir after the earthquake (see Figures 3.3i & 3.3j and 5.5h & 5.5i).

1.1g Photograph taken in the 1980s in Srinagar of an unreinforced masonry building without timber lacing, showing how soil subsidence is causing the structure to come apart.

1.1h View of partially demolished 18th- or 19th-century dwelling in Ahmedabad, Gujarat, with timber lacing like that found in Kashmir. This view shows the cross-pieces that tie the inside and outside runner beams laid into the masonry wall together. (A view of the peg holding one of these crosspieces is shown in Figure 2.3.3a.)

TAQ TIMBER BANDS

1.1i A model showing the typical "ladder bands" of timber runners and crosspieces found embedded in the masonry walls in traditional *taq* construction. At the floor levels, the joists run through the wall, and another pair of runner beams rests on top of the joists. At the window lintel level, shown in between, the timbers are tied together with cross pieces like those that can be seen in the Ahmedabad example in Figures 1.1h & 2.3.3a.

1.1j This detail of the corner of an abandoned house in Srinagar shows the historic masonry in taq *construction well. The hard-fired Maharaji bricks are only the external skin. The interior is random rubble of clay and low-fired bricks. The stout timber lacing consists of two layers of runner beams with the joists sandwiched in between. The ends of the joists can be seen between the runner beams in the facade on the right.*

With this amount of clay and random rubble, one would expect the earthquake performance of these structures not to be good, so the timber lacing undoubtedly is critical to the reported good performance. Had a section like this been broken off the corner of an unreinforced masonry building with no timbers to stop its progression, the collapsed section might have extended all the way to the roof.

1.1k Houses on the River Jhelum, Srinagar, to the north of First Bridge.

1.2a House of the Opus Craticium, *Herculaneum Archaeological Site, Italy.*

1.2 *Dhajji Dewari* Construction

Dhajji-dewari…has shown a marked resistance to earthquakes when compared to conventional unreinforced solid-wall masonry construction… The ability of the disparate materials, each of relatively low strength, to work together as a single system to resist catastrophic damage from the overwhelming forces of earthquakes is what makes these buildings so important.

Sanghamitra Basu
Indian Institute of Technology, Kharagpur[3]

Dhajji dewari is a variation of a mixed timber and masonry construction type found around the world in one form or another, both in earthquake and non-earthquake areas. While earthquakes may have contributed to its continued use in earthquake areas, timber and masonry infill frame construction probably evolved primarily because of its economic and efficient use of materials. Evidence of infill frame construction in ancient Rome emerged when archaeologists excavated the port town of Herculaneum that had been buried in a hot pyroclastic flow from Mount Vesuvius in AD 79. They found an entire two-storey half-timber house which is believed to be an example of what Vitruvius called *Opus Craticium* (Figure 1.2a).

After the fall of Rome, infill frame construction became widespread throughout Europe in earthquake as well as non-earthquake areas. In Britain it is called "half-timber", in France *colombage*, and in Germany *Fachwerk* (see Figure 1.3q). In earthquake-prone areas of Central America, Spanish construction was combined with local methods in what is today called *taquezal* or *bahareque*, in which a bamboo or split-lath enclosed basket between timber studs is filled with loose earth and stone. In South America, Peru is also seismically active, and the traditional construction with earthen plaster and sticks or reeds (wattle and daub) known as *quincha* that can be found there predated the Spanish conquest, after which it was adopted by the Spanish and continued in use almost until the present. Despite the ephemeral nature of the material, 5,000-year-old *quincha* construction has been unearthed at the Peruvian archaeological site Caral. All of this shows that for thousands of years, in many different cultures on different continents, people often have arrived at similar ways of building.

The term *dhajji dewari* comes from the Persian and literally means "patchwork quilt wall", which is an appropriate description for the construction to which it refers. The Persian name may provide a clue to Persian influence in the origins of this system of construction. It is also very similar to Turkish *hımış* construction, which was also common beyond the boundaries of Turkey, perhaps in part because of the widespread influence of the Ottoman Empire (see Figure 2b). *Dhajji dewari* consists of a complete timber frame that is integral with the masonry, which fills in the openings in the frame to form walls. The wall is commonly one-half brick in thickness, so that the timber and the masonry are flush on both sides. In the Vale of Kashmir, the infill is usually of brick made from fired or unfired clay. In the mountainous regions of Kashmir extending into Pakistan, the infill is commonly rubble stone.

Dhajji dewari frames are usually "platform" frames, meaning that each storey is framed separately on the one below. In *dhajji dewari*, the floor joists are sandwiched between the plates (see Section 3.6.1 for details on the structural importance of this). This framing distinguishes it from heavy timber frame construction which depends for its strength and stiffness on the posts which extend through more than one storey. In the first generation of sawn 2" x 4" (5 cm x 10 cm) stud "balloon frame" construction in the USA, the studs were extended through two storeys, and the floor joists rested on a timber that was framed into the studs, but this evolved into platform framing in the early 20th century, which is easier to build. Lacking continuity in its vertical timbers, platform frame construction depends for its stiffness on its enclosure membrane. In North American wood frame construction, this was first provided by diagonal sheathing, and now by plywood; in *dhajji* construction, it is the infill masonry.

While *dhajji dewari* construction evolved probably for similar economic and cultural reasons that led to the development of similar forms of construction around the world, its continued common use up until the present in Srinagar and elsewhere in the Vale of Kashmir most likely has been in response to the soft soils, and perhaps also to its good performance in earthquakes. *Dhajji dewari* construction is very effective in holding buildings together even when they are dramatically out of plumb (see Figure 4.2.1c for examples of this). In the mountain areas, where soft soils and related settlements of buildings are not a problem, its use continued probably because timber was available locally and the judicious use of timber reduced the amount of masonry work necessary, making for an economical way of building. Its observed good performance in past earthquakes may also have been a contributing factor, just as it is now again since the 2005 earthquake (see Sections 3.5 & 5.5). The panel sizes and configuration of *dhajji* frames vary considerably, yet the earthquake resistance of the system is reasonably consistent unless the panel sizes are unusually large and lack overburden weight.

Structurally, the timber frame and the masonry are integral with each other. In an earthquake, the house is dependent on them both, and their interaction with each other is an important part of their ability to resist collapse

1.2b & 1.2c A primitive form of dhajji *construction is still practised in Afghanistan, here seen in Kabul. Undressed poplar timbers are used together with unfired mud bricks. These two pictures show the progress of construction from open frame to the infilling of the frame, a process that is similar to that practised in Kashmir. Photos courtesy Aga Khan Trust for Culture, Afghanistan.*

Dhajji Timber Frame

1.2d Detail of *dhajji dewari* construction shows the thin masonry wall construction and the abundant use of timber. Historically, the system was economical because the total amount of masonry was radically reduced from what would have been needed for bearing wall construction, so long as the extra timber was available. Also, since the walls occupy a much smaller footprint, significantly more living space could be obtained on the same size lot. Notice how the floor joists are sandwiched between the timbers of the frame for each floor. Another view of this building can be seen in Figure 3.6.1e.

1.2e Building in Srinagar with first two storeys of taq *construction and top storey of* dhajji dewari.

in the tremors. Historically, the amount of wood used, and therefore the sizes of the masonry panels, varied considerably. There is evidence that walls with many smaller panels have performed better in earthquakes than those with fewer panels (see Section 4.2.4).

Dhajji dewari construction was frequently used for the upper storeys of buildings, with *taq* or unreinforced masonry construction on the lower floors (Figure 1.2e). Its use on the upper-floors is suitable for earthquakes because it is light, and it does provide an overburden weight that helps to hold the bearing wall masonry underneath it together. It was also sometimes used for the back and side walls of buildings with *taq* bearing wall construction on the front wall only. The earthquake stability of this combination does raise concerns because of the vastly different weight of the front wall compared to the rest of the exterior walls.

Dhajji dewari construction has continued in use to the present, although in recent decades reinforced concrete has become predominant in urban areas. As will be discussed in the last chapter, it is now worth revisiting both *dhajji* and *taq* construction, both for their earthquake-resistant attributes and for their potential economy and sustainable use of building materials.

Summary Comparison of *Taq* and *Dhajji Dewari* Construction

1.2f This illustration shows a model of *taq* timber lacing superimposed onto a photograph of part of a Srinagar *taq* house. The three timber bands illustrate two floors plus the window lintel band. This shows how the floor joists are sandwiched between double bands and line up with the single band along the opposing walls. The double beams on the inside of the lower floor band represent the extra beam that can sometimes be found when the masonry wall below is thicker.

1.2g This model shows a portion of a typical *dhajji dewari* house frame. The floor joists are sandwiched between the horizontal timber plates that form part of each storey of the frame, and the beams which support the joists at mid-span are integrated into the frames on the side.

There are often different variations on how the timbers are arranged, examples of which are shown on page 18. This model shows an example of an "X" brace on one side, and a single diagonal brace on the other. Some *dhajji* houses can be found which have no diagonals. Instead they rely on the brick infilling alone for lateral stability. The top detail illustrates how the top floor joists extend out to form a cornice above which the roof rafters are placed.

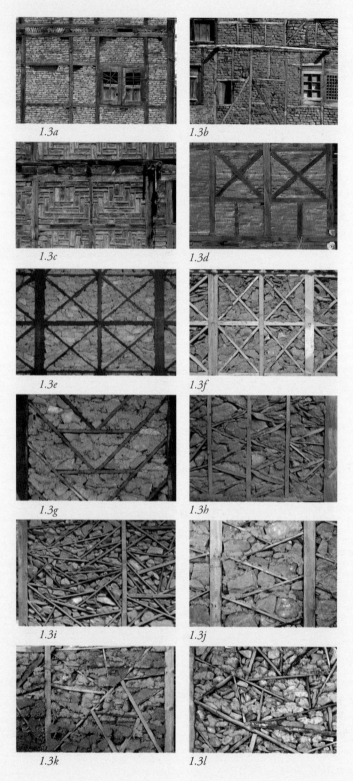

1.3a

1.3b

1.3c

1.3d

1.3e

1.3f

1.3g

1.3h

1.3i

1.3j

1.3k

1.3l

1.3a-d The masonry infill in Srinagar and the Vale of Kashmir is traditionally of brick with larger panels than in the mountain areas on both sides of the Line of Control.

1.3e-l In rural mountain areas of Indian and Pakistan Administered Kashmir, the infill is of stone, with smaller panels. Most of these patterns have their own individuality, based on a combination of the builder's skill and craft traditions handed down locally, while the more random layouts are also a function of using variable lengths of short timbers without having to cut them smaller, thus eliminating waste.

Photos 1.3e-g, 1.3j, and 1.3l by Maggie Stephenson, UN-HABITAT.

1.3m This rural farmer living remote from any village in the mountains between Bagh and Muzzafarabad, in Pakistan Administered Kashmir, was photographed in 2006 while single-handedly reconstructing his house in dhajji *construction after the earthquake. Most of the wood for the project was salvaged from his prior home that had collapsed. The random pattern in Figure 1.3k is his creation.*

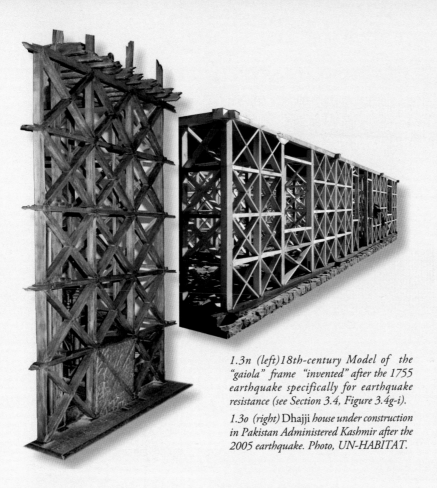

1.3p

1.3q

1.3n (left)18th-century Model of the "gaiola" frame "invented" after the 1755 earthquake specifically for earthquake resistance (see Section 3.4, Figure 3.4g-i).

1.3o (right) Dhajji house under construction in Pakistan Administered Kashmir after the 2005 earthquake. Photo, UN-HABITAT.

1.3r

1.3s

1.3 CONTINUITIES ACROSS TIME AND SPACE

Dhajji construction is not unique to Kashmir, as similar construction with different names can be found in earthquake areas and non-earthquake areas alike around the world, and across millennia of time. One rather remarkable convergence can be found with the comparison with the "*gaiola*" developed in Lisbon after the 1755 earthquake (See Section 3.4 for a full description) with the timber frames now being built for some of the more refined examples of new post-earthquake housing in rural Pakistan Administered Kashmir.

There is no research that demonstrates that one *dhajji* pattern is better than another, and some patterns lack diagonal bracing elements, relying on the masonry to provide the bracing. The random patterns probably result from the economics of using available random lengths of wood in the most efficient way possible. In fact, the quilting from which it gets the name '*dhajji*' is itself produced from the reuse of scraps and small pieces of cloth.

1.3t

1.3p 15th-century half-timber in Britain
1.3q 18th-century Fachwerk *in Germany.*
1.3r 18th-century gaiola *in Portugal.*
1.3s 19th-century French colombage *in New Orleans, USA.*
1.3t Mid-20th-century decorative hımış *in Turkey.*
1.3u Post-1999 earthquake hımış *in Düzce Turkey.*

1.3u

2a The unreinforced masonry school in the government cantonment in Uri completely collapsed in the 2005 Kashmir earthquake. Ramazan Mubarak written on the undamaged blackboard means "Blessed Ramazan" (more commonly transliterated "Ramadan"). The October 8, 2005 earthquake occurred during the Ramadan period.

An Introduction to Structural Concepts of
Earthquake Safety

The greatest danger to historic buildings comes from engineers who are unaware of their unique values and apply the Codes literally... It can be said with some justice, that many a historic building has the option of being destroyed by the Codes or by the next earthquake.

Sir Bernard M. Feilden
Conservation of Historic Buildings, 1982

Earthquakes are unique among natural hazards in that they occur without warning and thus can place at risk people who in other natural hazards would be able to retreat to a safe location. This is why each structure used by people in an earthquake-prone region must be able to avoid collapse in an earthquake that may strike at any moment.

Most building codes in most countries subject to high seismicity, including India and Pakistan, are based on the objective of "life safety" and thus are meant only as minimum standards to prevent collapse. Except for highly important structures which must remain functional during and immediately after an earthquake, such as nuclear power stations, fire stations, hospitals, etc., codes are not designed to prevent damage, even severe structural damage, in large earthquakes.

The evaluation of earthquake risk, both to individual buildings and for a community as a whole, is a complex task. First, the potential magnitude and frequency of future earthquakes at the site must be estimated by seismologists. Site-specific soil conditions must be determined in order to estimate the amplification of the shaking at the site.

Next, based on this data, new and existing building design and building materials and construction quality must be evaluated with regard to their ability to withstand the expected shaking. Then plans must be drawn up to upgrade unsafe buildings and to eliminate specific hazards, such as unbraced parapets and other non-structural building components that can fall onto people below. The process of assessing risk and planning for earthquake upgrading calls on a variety of disciplines such as

2b Turkish timber and masonry infill-frame construction, known in Turkish as hımış, *photographed one month after the 1999 Kocaeli earthquake, in the heart of the damage district in Gölcük. Despite the fact that many nearby reinforced concrete buildings collapsed, this and the neighbouring houses of this construction mostly suffered very little damage, notwithstanding the rudimentary level of craftsmanship evident in their construction.*

structural engineering, seismology, public administration, risk analysis, statistics, economics, building conservation technology, and materials testing and inspection, to name just a few.

What may seem obvious at first glance may become less obvious upon more detailed examination. It is easy for inspectors undertaking surveys to condemn traditional buildings as hazardous because of their age, their ragged looks, or their archaic structural systems. Upon closer inspection, on the basis of actual earthquake experience, many of these traditional buildings will be found to be less risky than some newer structures which look neat and safe but are built with reinforced concrete frames with hidden deficiencies. Also, seemingly solid masonry walls can turn out to have rubble cores and/or to fail to have been securely connected to the floors and roof. Poor construction, even of recent vintage, can be hidden beneath a smooth layer of plaster.

This counterintuitive contrast between resistant and vulnerable construction proved to be the case in the great Kashmir earthquake of 1885 when traditional vernacular buildings in Srinagar, which looked "tumbledown and dilapidated", proved to be resistant to collapse, while the seemingly well-constructed Maharaja's palace collapsed (Neve, 1913).

2.1.1a Collapsed residential blocks in Adapazari, Turkey, after the 1999 Kocaeli earthquake. Rows upon rows of these blocks suffered pancake collapse and others, such as these, sank into the liquefied soil or collapsed because of weak ground storeys. In one area in this city, only the historic stone mosque remained standing amidst the ruins of pancaked reinforced concrete apartment buildings.

2.1.1b Tipped over by the 1999 Kocaeli earthquake in Turkey, this building in Adapazari demonstrates the effects that extremely soft soils subject to "liquefaction" can have on structures. In this case, the reinforced concrete structure was strong enough to remain together, a testament to a well-constructed superstructure, but its narrow base and foundation made it subject to tipping over when the earthquake vibration caused it to rock from side to side while at the same time causing the soft water-laden soil to liquefy.

2.1 How Earthquakes Affect Buildings

(NOTE: The words in bold are explained in this chapter. For other technical terms, such as those shown in this chapter in quotation marks, please consult Appendix 2: Glossary of Technical Terms.)

2.1.1 Soil Characteristics and Resonance: Buildings can also be severely damaged when the soils that support the foundations vibrate, shift, sink, slide, or liquefy, in response to an earthquake, even when the earthquake's epicentre is distant from the site. The two major types of earthquake waves are (1) the initial "primary" or "P" waves, which are compression waves that are short and jolting, and (2) the more damaging "secondary" or "S" waves, which are shear waves that cause a back-and-forth rolling motion. The waves change frequency and amplitude as they travel through the soils.

Interestingly, shorter and stiffer buildings resonate more with the short-period motion that occurs near an earthquake's epicentre, while taller buildings, which have a longer period, resonate more with the longer period motion characteristic of more distant earthquakes. Sites on or near bedrock, such as in the mountain areas of Kashmir, transmit short-period waves, which are more devastating to short and stiff buildings which have a period of vibration similar to that of the earthquake. This is called "resonance". This phenomenon contributed to the extensive damage to low-rise unreinforced masonry and reinforced concrete buildings in the epicentral region of the 2005 Kashmir earthquake in Pakistan and the western part of Indian Kashmir.

Soft, loose soils, such as those found in Srinagar and throughout the Vale of Kashmir, tend to amplify the longer-period ground motion characteristic of earthquakes with epicentres distant from the site. In many cases this effect can even make the earthquake last longer and may accentuate building damage. In the 2005 earthquake, with the soft soils, one would have expected the damage to the more flexible traditional *taq* and *dhajji dewari* buildings in Srinagar and Baramulla to have been more severe than it was, but this is where the energy dissipation inherent to both of these systems becomes a life-saver. The buildings are flexible but not very elastic. It is the **damping** from their inelastic behaviour that reduces their resonant response to the ground motion (see Section 2.2.3).

Site response was a major problem in the 1985 Mexico City earthquake, in which the total collapse of numerous modern tall buildings caused many fatalities. The earthquake occurred 350 km (220 miles) from the city, but very soft soils beneath a 6 km square area in the city amplified the ground shaking, and many tall buildings – including even one 21-storey building – collapsed because they resonated with this long-period shaking, but lacked sufficient **damping** to decrease this resonance.[5]

2.1.1c An example of resonance: The famous "Shaking Minarets" of the Sidi Bashir Mosque (ca.1454) in Ahmedabad. If a sway is initiated in one of them, the other begins to sway in resonance with it without any detectable movement in the connecting passage. The 2001 Gujarat earthquake caused damage, some of which can be seen in the photograph where the stones are slightly separated on the first-level circular balcony.

British Infantryman Robert Melville Grindlay wrote in 1809 "the most remarkable circumstance attached to this building is the vibration which is produced in the minarets... by a slight exertion of force at the arch in the upper gallery."[6] The British were so mystified by this phenomenon that they dismantled part of one minaret on the Raj Bibi Mosque, another Ahmedabad mosque that was subject to the same behaviour. This minaret remains dismantled to this day because they did not know how to put it back together.[7]

2.1.2a A modern reinforced concrete frame with masonry infill structure in Italy, showing in-plane failure of the ground-floor masonry wall resulting in diagonal tension "X" cracks characteristic of shear failure of the masonry infill.

2.1.2b Out-of-plane wall failure in Italy of an unreinforced masonry building resulted in the façade falling away from the building into the street.

2.1.2c This building in Italy shows a combination of both in-plane and out-of-plane failure of walls. The side wall failed primarily (but not exclusively) from in-plane forces as shown by the diagonal tension cracks visible in the remaining part of the wall, and the rear wall appears to have collapsed primarily from out-of-plane forces.

Optimally, structures should not be located in areas with poor site conditions, but in Kashmir as in many other places in the world, human needs for water, water-borne transportation, and fertile ground have led to the existing settlement patterns, which for both economic and cultural reasons must be maintained and allowed to grow. In places like Kashmir, these sites have required, and continue to require, creative ways to respond to the problems presented by soft soils such as differential settlement and increased earthquake risk.

2.1.2 IN-PLANE VS. OUT-OF-PLANE DIRECTION: Since earthquake forces are transmitted through the soil as waves, the forces on a building are directional. Buildings must be designed to resist collapse from earthquake forces from any direction because it is not possible to predict what direction the largest vibrations will be coming from. However, the structural requirements are different depending on the direction of the waves. That is why buildings with identical structural systems and architectural configurations but on intersecting streets may suffer widely different levels of damage in an earthquake.

For example, those walls which are parallel to the direction of the movement of the earthquake's waves across the ground ("in-plane walls") must resist the in-plane earthquake loads, whereas those walls which are perpendicular to the direction ("out-of-plane") must resist the forces normal (perpendicular) to the plane of the walls. This can mean that the out-of-plane masonry walls can fail by **overturning**, or, if thin and tall, by folding in the middle and collapsing, whereas the in-plane walls may develop large diagonal "X" cracks (known as "diagonal tension cracks" or "shear cracks") and collapse in place (see Figures 2.1.2a and 2.2.1a). Thus, every wall and the building as a whole must be designed to respond to both of these force directions.

For unreinforced masonry bearing wall buildings, the ratio of the floor-to-floor height of a wall to its thickness is important. In the code developed in the United States that applies to the mitigation of the earthquake hazard in existing unreinforced masonry buildings, height-to-thickness ratio maximums are specified, above which walls are required to be braced. For the highest seismic zone, these ratios range from 9 to 15 depending on the location of the wall (UCBC, 1994).

2.2 HOW BUILDINGS RESIST EARTHQUAKES

2.2.1 LOAD PATH: The structural systems of buildings are primarily designed to resist static gravity forces. The roof and floor systems carry these vertical forces to the supporting beams, which then transfer them to the columns and bearing walls and thus to the foundation and the supporting soil. This sequence of transfer of inertial forces from the roof and floors to the walls and columns, and from there down to the soil, is known as a "load path". Different structural systems use different load-bearing elements and rely upon different load paths. Both wind and earthquakes introduce horizontal ("lateral") forces into a structure which must be resisted by the structural system. When analysing the capacity of a given building to adequately resist collapse in an earthquake, an engineer or architect must calculate the lateral loads along all parts of the building's load path, and then calculate the capacity of the components of the structural system along that same path to resist those forces. (Foundations are an important part of the load path, but are not the subject of this volume.)[8]

Earthquake-resistant systems are designed to resist forces that result mainly from horizontal ground motion. The vertical forces from earthquakes are not usually considered in the design process because reserve capacity of the floor and roof systems designed to meet maximum dead and live loads is generally considered to be sufficiently large enough to resist the vertical earthquake loads. Unless the floor spans are very large or there are long cantilevers, the vertical dynamic characteristics of most buildings are such that there is very little dynamic response to vertical ground motion.

(Near the epicentre of large earthquakes, vertical forces can be close to or exceed the force of gravity, and thus have been known to contribute significantly to damage of buildings, as for example, in the 2003 Bam, Iran earthquake. However, the risk of this happening is significantly less than the risk of damage from horizontal shaking because the odds of being directly over the hypocentre of a relatively shallow earthquake are significantly less than the risk is of being within the overall damage district of an earthquake.)

The earthquake forces on a building result in tension, compression, shear, bending, or torsion forces on the building's structural system (Figure 2.2.1b). It is a building's "lateral-force-resisting system" (or "earthquake system") that is called upon to resist the effects of horizontal earthquake forces which can come from any direction. The failure of structural components or connections along the load path on which the building depends for stability can lead to partial or total structural collapse.

2.2.2 STRENGTH AND STIFFNESS VS. FLEXIBILITY:
Strength is the property of an element to resist force, while stiffness resists displacement. When two elements of different stiffness are forced to deflect the same

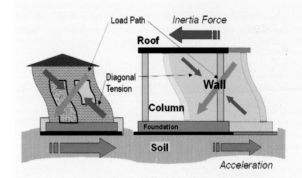

2.2.1a Inertia Force and Load Path: The "load" is the weight of the building itself ("dead load") and the weight of the moveable contents ("live load"). When the building moves in an earthquake, the building's "earthquake load" is a product of the earthquake's acceleration of the ground and the inertia of the static weight of the structure. This is represented as a percentage of gravity (g). The "load path" through the wall parallel to the ground vibrations is represented by diagonal "struts" or lines of force through this wall. The deformation of the wall results in tension forces in the opposite direction, which are the cause of the diagonal cracks found on damaged walls. The system reverses direction with each earthquake vibration, so "X" cracks are often produced from the diagonal tension forces. Diagram courtesy of Dept. of Engineering, IIT-Kanpur.

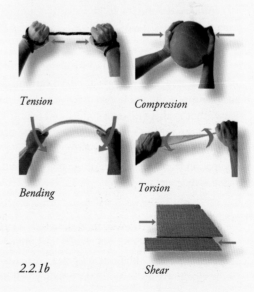

Tension   *Compression*

Bending   *Torsion*

2.2.1b   *Shear*

amount, the stiffer element will carry a larger percentage of the total force because it takes more force to deflect it. When stiff concrete (or masonry) elements are combined with more flexible steel or wood elements, the concrete (or masonry) takes most of the total force.

The stiffer a structure is, the larger the earthquake forces are that act on it and thus the stronger it must be to resist those forces. In many cases modern earthquake systems, such as reinforced concrete **shearwalls**, are specifically designed to minimize the amount of relative lateral movement and consequential damage by stiffening and strengthening the structure. These construction systems are stiff and derive their resistance from their strength rather than their ability to be flexible and give with the earthquake vibrations.

By contrast, a flexible approach reduces the earthquake forces that the structure must resist by allowing for larger lateral movement. A flexible structure can be at risk if its natural period of vibration is close to the frequency of the earthquake's ground motion at the site such that it resonates with the ground motion sufficiently to sway so much that it collapses. This risk can be mitigated by **damping** its vibration. **Damping** increases when the building's

2.2.2a & b The structure on the left is rigid, and the one on the right is flexible. In 2.2.2b the combination of the clay, which is visible between the woven twigs, and the tightly woven wood produces frictional damping when the structure is vibrated.

structural system begins to go inelastic. In other words, it can be the onset of damage to a building that can be essential to its resistance to collapse, so long as the damage is not concentrated in elements critical to the stability of the structure.

The two main traditional construction systems of urban Kashmir, *taq* and *dhajji dewari* and similar construction in Gujarat and Turkey, have survived earthquakes by having sufficient flexibility to allow movement to occur in the masonry walls, but not so much as to lead to instability of the buildings. A large amount of frictional **damping** is inherent in these types of timber and masonry systems. This damping results from friction along the interfaces between the timber and masonry and in the mortar joints of the masonry itself. Both *taq* and *dhajji dewari* are systems designed to begin to yield inelastically almost at the onset of shaking. This then mobilizes the energy dissipation from the friction of the masonry and timber together, while the timber keeps the masonry walls from being torn apart.

2.2.3 DAMPING: When a tuning fork is struck, it vibrates back and forth at a certain rate. This rate is known as its fundamental period. All objects, including buildings, have their own unique fundamental period of vibration. If the earthquake's vibrations at a given site matches the fundamental period of a building at that site, the amplitude of the vibrations at the top of the building will be much greater than at its base. This can lead to extensive damage or sudden collapse, unless this energy is "damped" out (dissipated) by the inelastic behaviour of the building's structural and non-structural components prior to the critical failure of its vertical load bearing system. For a bolted steel frame building, this damping is the result of a combination of molecular damping from the bending of steel members beyond their elastic limit and frictional damping from friction in the bolted connections. In traditional timber and masonry structures, damping is derived mainly from the friction of the components cracking and then rubbing against each other as the materials that make up the walls and floors shift and slide.

Damping in building structures is a combination of "viscous damping" and "hysteretic damping". Viscous damping is a linear elastic behaviour from friction of materials moving or bending, while hysteretic damping is the damping from the non-linear behaviour of the materials and system. Damping diminishes a building's resonance with an earthquake by dissipating the energy in the same way that a shock absorber (a viscous damper) in a car dampens the car's vibrations from bumps in the road. After the onset of damage, damping is imparted to a building by the cracking and inelastic movement of its structural components (hysteretic damping). Non-linear behaviour is the permanent yielding of the components of the structure, i.e. it is, by definition, damage to the structure.

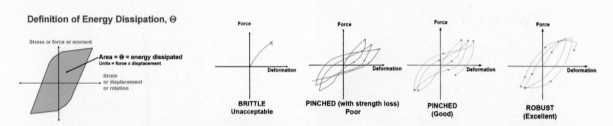

2.2.3a Hysteretic (non-linear) behaviour under seismic excitations. Each loop represents one complete cycle of motion back and forth, and the area inside each loop represents the total energy that is dissipated over each cycle. The variations in the hysteresis diagrams for different materials and structural systems illustrate the kind of behaviour that is either deficient or beneficial in an earthquake. These illustrations help to explain the hysteresis diagram produced from tests of the 18th-century Portuguese Pombalino system shown in Figure 3.4i. In that case, the hysteresis diagram was generated by tests in a lab to document the inelastic response of a building's structural system provided by the walls of timber and infill masonry. (Source: FEMA 451.)

"Design-level" earthquake forces are simply too large for all but the most important and costly buildings, such as nuclear power plants, to be designed to remain elastic. Thus controlled damage is necessary for affordable design because the damping it produces is what will avoid deformation of the structure that is sufficiently large enough to cause collapse (see Figure 4b for an example of controlled damage in a traditional structure in an earthquake). This behaviour underlies the effectiveness of traditional Kashmiri *taq* and *dhajji dewari*, in the same way that it is fundamental to the more advanced modern systems which utilize specially designed connections or mechanical dampers to dissipate energy and reduce a building's response to earthquakes (as for example in Figure 2.2.3b).

2.2.4 DUCTILITY: Under normal conditions, a building bends when a force is applied and returns to its original shape when the force is removed. This is known as "elastic" behaviour. However, as described above, extreme earthquake forces may generate "inelastic" deformations in which the component is distorted sufficiently to result in a permanent deformation after the force is removed. Ductility is the property of a material or system that allows it to sustain such inelastic deformations without breaking. In engineering terms, this means it can undergo repeated

2.2.3b This is an example of a shock absorber for a new building that is intended to function in an earthquake like a car shock absorber functions on a bumpy road, except that it is positioned to dampen horizontal, rather than vertical, motion. This manufactured earthquake damper is shown being installed between the first-floor and the foundations of a house designed by the author in the USA as part of a high-damped base-isolation system. (More information about this project is available at www.conservationtech.com.)

2.2.4a Both the stones themselves, and the unconfined masonry pier as a whole of the Bhuj Darbargarh demonstrated brittle failure during the 2001 Gujarat, India earthquake.

2.2.4b The ductility of steel has allowed this column to deform without breaking during the 2003 Bam, Iran Earthquake. Many steel buildings did collapse in this particular earthquake, but the collapses were primarily because of poor quality welding, which caused brittle failure of the connections.

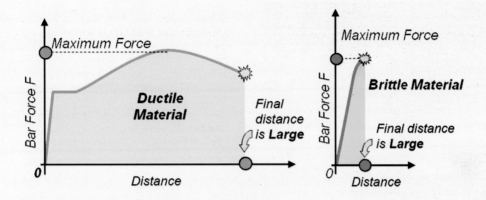

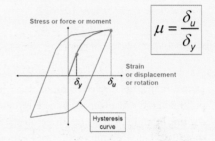

$$\mu = \frac{\delta_u}{\delta_y}$$

2.2.4c **Ductility** is an attribute that applies to the behaviour of a single material, and also a system composed of many materials, such as a building structure. An adequate design is accomplished when a structure is dimensioned and detailed in such a way that the local **ductility** demands (energy dissipation demands) are smaller than their corresponding capacities.

Ductility, the property of being able to deform without breaking, is defined by the inelastic component of the strain. (The curved shape of the line on the hysteresis diagram explains the derivation of the term "non-linear"). The diagram on the right represents a cyclical motion, as caused by an earthquake, and the two above represent a force applied in a single direction. *Above: Dept. of Engineering, IIT-Kanpur; Right: FEMA 451.*

and reversing inelastic deflection while maintaining a substantial portion of its initial maximum load-carrying capacity. Building materials such as steel are highly ductile, whereas both concrete (without adequate steel reinforcement) and masonry are not ductile.

Building elements constructed with ductile detailing have a "reserve capacity" to resist earthquake overloads. Generally speaking, the greater the ductility without significant loss of strength, the better. Therefore, buildings constructed of ductile components such as steel (including the steel within well-constructed reinforced concrete) tend to withstand earthquakes much better than those constructed of brittle materials such as unreinforced masonry. However, it is also possible for a "system", i.e. the combination of materials in a structural system, to behave as if it is ductile even though some of the individual materials within it are relatively brittle. This is why reinforced concrete can be considered to be ductile if it contains enough well-placed steel, even though the Portland cement concrete itself is brittle. This is also an attribute of the combined timber and masonry systems in Kashmiri traditional buildings which gives them their earthquake resistance. The brittle behaviour of the masonry leads to initial cracking and shifting in an earthquake, but the confinement and reinforcement of the masonry by the timber in both systems works to maintain the strength and stability of the buildings.

STRENGTH versus CAPACITY: An explanation of why strength
ALONE DOES NOT DETERMINE CAPACITY TO WITHSTAND EARTHQUAKES

2.2.4d This stress/strain diagram shows the relative differences between the behaviour of different materials and construction systems. This diagram is not intended to predict the behaviour of any specific structures in earthquakes. Instead it is a conceptual diagram intended to illustrate the concepts rather than the result of any specific tests. The nature and slope of the curves is derived from research papers cited in the bibliography (see Paikara, 2005 & 2006, and Santos, 1997).

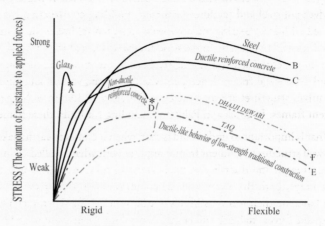

Line A represents the behaviour of glass: the line rises steeply indicating very little elastic range and then stops at the point when the glass breaks. Line B represents the behaviour of a properly riveted, bolted or welded steel structure: the line would normally continue off the chart before it breaks, representing the fact that steel has a great deal of **ductility**. Line C is for a reinforced concrete moment frame that is designed and constructed with ductile detailing in the beam/column joints. Line D represents non-ductile reinforced concrete moment frame construction. Non-ductile reinforced concrete is construction that lacks the volume and proper placement of steel reinforcing necessary to give it ductile behaviour in the inelastic range. Very often, in addition to lacking ductile design, many buildings in earthquake areas in India and other countries also suffer from poor construction practice such as poorly mixed, inadequately hydrated and/or improperly placed concrete, any of which can be the cause of collapse in even a moderate earthquake.

Lines E and F represent *taq* and *dhajji dewari* respectively. The initial elastic strength of these systems is much lower than that of reinforced concrete, but the inelastic behaviour of the systems is favorable to the survival of the traditional buildings in earthquakes because of the ability of these buildings to undergo large inelastic deformations without coming apart. With *taq* (line E), the post-elastic capacity in earthquakes has often been found to be substantial because of the presence of the timbers which continue to hold the masonry together after it begins to crack in-plane, and then shift and slide. The timbers also serve to block the propagation of large shear cracks in the masonry and resist the out-of-plane forces, thus helping to maintain the stability of the wall and increase the frictional **damping** from the myriad of smaller cracks that open instead.

In the case of *dhajji dewari* (line F), the initial shallow slope represents the **flexibility** of the timber frame, while the kink in the line represents the point at which the infill masonry begins to be compressed against the confining frame of timber as the deformation of the frame increases. The ultimate strength of the system is thus determined by the crushing strength of the confined masonry, mortar, and timbers in combination. The gradual slope to the right represents the continued "working" of the confined masonry panels in their frames, even as they slowly degrade.

Unreinforced masonry (URM) has not been represented on this chart because such construction can vary in performance over such a wide range, from rubble stone in mud mortar which tends to collapse very quickly, to well dressed horizontally bedded ashlar which has demonstrated its ability to survive earthquakes, such as the 1999 earthquakes in Turkey where unreinforced masonry mosques with their stone minarets survived intact, while scores of modern reinforced concrete buildings collapsed around them. In general, though, URM lacks the ductile-like behaviour of *taq* and *dhajji dewari* because of the absence of the timber reinforcement.

2.3 Architectural and Structural Design for Earthquakes

2.3.1 The Emergence of the Skeletal Structural Frame for Modern Buildings:
With the advent of steel and reinforced concrete, building construction has shifted from a preponderance of buildings designed with load-bearing masonry walls to structural frames, with the enclosure walls designed to be supported by the frames. Historically, frame structures were limited to non-rigid frames of timber and bamboo, but with the advent of structural steel (including steel-reinforced concrete), the limitations of size were eliminated, giving birth to modern skeletal frame skyscrapers where both the vertical and lateral loads are carried by a rigid frame. Now most urban structures are constructed with structural frames with rigid beam/column connections, referred to as moment frames. In India and Pakistan, including Kashmir, these frames are usually of reinforced concrete.

What is important to understand is that, when it comes to earthquakes, moment frames respond differently than solid-wall structures. Moment frames respond with what is called "frame action", wherein lateral forces are resisted by the strength and **ductility** of the beams and columns and their joints. When the frame is deflected within its elastic range, both the beams and the columns develop complex "S" shape bends, with all members sharing the forces based on their relative stiffness (see Section 2.2.2). When such rigid frames are forced to yield inelastically by earthquake forces, stability will largely depend on the amount of residual strength and hysteretic **damping** that is available in the yielding portions of the structure.

In addition to moment frames, modern frame construction includes two other types: frames with **shearwalls** (frames with strong solid walls at specific locations extending from foundation to roof), and braced frames (frames with diagonal braces). Both of these systems generally have had a better record in earthquakes than have moment frames, as they are less sensitive to faults in design and construction, and less negatively impacted by the weight and stiffness of non-structural infill and enclosure walls.

The typical reinforced concrete frame building found in India and Pakistan, including Kashmir, is a moment frame because this system requires less steel and cement than the other two systems and is therefore usually more

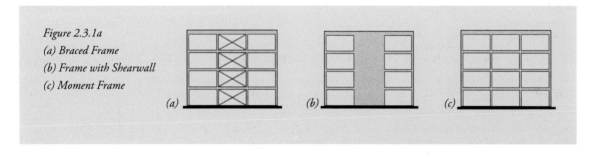

Figure 2.3.1a
(a) Braced Frame
(b) Frame with Shearwall
(c) Moment Frame

(a)   *(b)*   *(c)*

2.3.1b The building on the right shows a typical multi-storey reinforced concrete moment frame building. It was under construction in Bhuj, Gujarat at the time of the 2001 earthquake. Because it was incomplete, with no masonry infilling to interfere with the frame action of its moment frame, it survived the earthquake despite poor construction quality. Its neighbour, as well as many other occupied reinforced concrete buildings in the neighbourhood, was not so lucky. It suffered soft-storey collapse (see Section 2.3.4).

economical. Unfortunately for most of these reinforced concrete moment frames, when completed as buildings with infill masonry walls, the frame action is corrupted by the addition of the infill walls. These walls create zones of stiffness, forcing the frame to undergo stresses and localized extreme deformations that it was never designed to resist (for a comparative example, see Figure 2.3.1b). Of the three types of modern reinforced concrete structural systems described here, the moment frame with masonry infill walls has proven to be the most prone to earthquake collapse because the proper detailing and construction of beam/column joints with sufficient strength and **ductility** to survive a design-level earthquake requires extensive knowledge, training and skill.

An alternative reinforced concrete construction system of growing popularity is "confined masonry", which is a variation on the shearwall type, but which requires only about the same amount of cement and steel as moment frames. This system has a much better track record in earthquakes because the infill masonry is utilized as an intentional and integral part of the lateral resisting system. To date, it has generally only been used for low-rise buildings. (For a description and an illustration, see Appendix 2: Glossary of Technical Terms.) UN-HABITAT in Pakistan has reported success in getting people to choose it after the 2005 Kashmir earthquake.

2.3.2 SHEARWALLS VS. CROSSWALLS: A secondary, but important, set of structural components are crosswalls, which are interior walls and partitions that are not necessarily continuous to the foundations, but which are attached securely to the top side of a floor "diaphragm" and to the underside of the floor above, and which are stiff and strong enough to resist the independent movement of the two connected diaphragms (see also **redundancy**). Such internal partitions are usually not designed to be shearwalls. Shearwalls are constructed to collect most of the earthquake forces, which are then resisted by strength and controlled damage and deformation. They are usually engineered as part of the primary structural system of a building, and they serve best if they are placed symmetrically as far apart as possible, resting on strong foundations and running continuously from ground level to roof without any offsets, whereas crosswalls are often not lined up floor-to-floor.

Crosswalls are expected to yield in an earthquake, but not collapse. In yielding, they help the building resist collapse by dissipating energy while maintaining substantial residual strength. Crosswalls need only be securely connected to two diaphragms (the floor below and above), and there must be a similar number of such walls relatively symmetrically arranged on every floor, including the ground-floor, in each principal direction. The interior partitions of *dhajji* construction in *taq* and *dhajji dewari* buildings are crosswalls. (The author, with support from engineers from the USA, Italy and Turkey, has proposed a seismic hazard mitigation system, called Armature Crosswalls, that

2.3.3a *Ruins of an 18th-century house in Ahmedabad showing the peg that holds the cross-timber to the timber runner beams laid into the walls.*

2.3.3b *Modern galvanized steel anchor plates for lightweight timber construction. Source: Simpson Strong-tie Company Inc.*

is based on *dhajji dewari*, to be used to reduce the risk of collapse in modern buildings of reinforced concrete moment frame and infill wall construction. Information on this can be found in Langenbach, 2006b and at www.conservationtech.com)

2.3.3 CONNECTIONS: Strong building connections allow forces to be transferred between vertical and horizontal building elements. Strong connections also increase overall structural strength and stiffness by allowing all of the building elements to act together as a unit. Inadequate connections represent a weak point in the **load path** of the building and are a common cause of earthquake damage and collapse. This principle is true for modern and traditional buildings alike. The earthquake effectiveness of connections may not only be a function of the strength of the connection itself, but also of the characteristics of the surrounding materials. In masonry, for example, it is also a function of overburden weight and friction. In traditional construction, before the 20th-century introduction of steel bolts and gusset plates, the overburden weight of the masonry contributed to holding the notched, nailed or pegged timbers, which in turn kept the masonry from spreading and coming apart (Figure 2.3.3a). In modern timber frame construction, the need for mortises and notches, which reduce the strength of the timber, has been eliminated with the manufacture of surface mounted steel hardware (Figure 2.3.3b). Steel frame construction at first depended on rivets where today bolts and welding are used. Reinforced concrete depends on the careful placement of steel bars with sufficient overlaps, together with the proper placement and compaction of the concrete surrounding the steel cage.

2.3.4 WEIGHT AND STIFFNESS DISTRIBUTION:

(1) SOFT STOREY/ WEAK STOREY: Buildings with shops or open parking areas at the ground-floor level with fewer interior partitions and exterior walls than above have a "soft" and "weak" storey (Figures 2.3.4a & b). This is extremely dangerous in an earthquake, as it can lead to the sudden collapse of the entire building. This is known as a "pancake" collapse, because the walls fall away, leaving the floors to pile on top of each other like a stack of pancakes (*chapatis*). Soft storeys were most responsible for the collapse of many reinforced concrete buildings in the 2001 Gujarat earthquake. Although the soft/ weak storey problem has become common with modern structures with shops or parking at the ground level, historic buildings with insufficient interior or exterior walls at the ground-floor are not immune (Figure 2.3.4c) (see "Basic Structural Engineer's Checklist" on page 89 for guidelines on the correction of this condition).

(2) TORSION: Buildings with an asymmetrical layout in plan of stiffer and stronger elements, such as the staircase or solid walls, can be vulnerable to torsion, as the earthquake will cause the structure to twist around the stiffer elements, causing excessive damage or

2.3.4a This reinforced concrete apartment block in Ahmedabad shows an extreme case of a soft storey at base of structure. Many buildings with this configuration collapsed in the 2001 Gujarat earthquake.

2.3.4b Ground-floor of a 6-storey building in Turkey close to collapse as a result of the 1999 Kocaeli earthquake.

2.3.4c Similar problem with a timber frame house in Adapazari in the same earthquake. Remarkably, the ground -floor collapse of this house was arrested by a traffic sign on a steel pole visible on the left.

2.3.4d An example of short column failure in Bhuj caused by the 2001 Gujarat earthquake. The short stubs of the columns that are above the infill-masonry wall failed in shear because the wall prevents the columns from bending along their whole length. Photo by IIT-Kanpur Professor C.V.R. Murty for EERI.

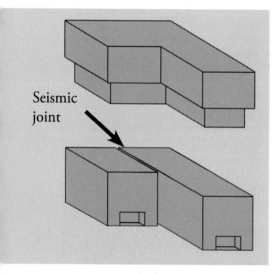

Seismic joint

2.3.5a The top illustration represents a three-storey building with both a vertical and horizontal complexity that can make it more vulnerable to damage in earthquakes than the building shown below where the upper walls come directly to the foundations, and a seismic joint separates the superstructure into regular shapes without the re-entrant corner. It is also important to avoid torsion by placing internal shearwalls, elevators, and stairwells symmetrically around the central axis of each part.

precipitating collapse from **overturning** of the columns at the more flexible end of the structure (see Figure 4.3.1g). While putting a staircase in the corner of a multi-storey building may be more efficient, it is potentially dangerous in an earthquake unless some other stiff and strong element is installed in an opposite corner to make the building structurally symmetrical, or the stairs themselves are designed to slide at their base so they cannot behave as unintended struts.

(3) OVERTURNING: Buildings that are wide at their base and have most of their weight distributed to their lowest floors generally perform better in earthquakes than do taller, slender buildings, especially if top-heavy (see Figure 2.1.1b). (Skyscrapers, however, are usually less at risk from earthquakes than mid-rise buildings because of the inherent strength needed to resist wind and gravity loads, and because their long period prevents them from resonating (except if located on a very deep soft soil layer as, for example, in Mexico City, where during the 1985 earthquake a 20-storey building collapsed).[9]

(4) SHORT COLUMN FAILURE: One phenomenon characteristic of reinforced concrete infill frame construction that has often led to whole building collapses (of schools in particular) is "short column failure". This occurs when a frame structure is deflected in an earthquake, but certain of its columns are partially confined by walls leaving a short unconfined part that cannot bend, while those columns not trapped between these short wall elements are allowed to bend over their entire length. This then results in the shear failure of the "captive" columns across an entire column line – such as in a school with a band of windows. These column failures can precipitate the collapse of the building (see Figure 2.3.4d).

2.3.5 BUILDING CONFIGURATION: Square or rectangular buildings with floor plans with symmetrically placed lateral-force-resisting elements perform better in earthquakes than buildings composed of irregular shapes. Re-entrant corners and unsupported overhangs are likely to increase vulnerability to damage (see Figure 2.3.5a).

2.3.6 REDUNDANCY: The term "redundant" has a positive meaning in earthquake engineering. Redundant elements serve as a kind of insurance policy against collapse by providing additional structural tissue that can take the loads off primary structural elements in an earthquake. It is very beneficial for a building to have an appropriate level of redundancy, so that any localized failure of a few elements of the system will not result in overall instability or collapse of the building. This is one of the important features of residential buildings, including those of traditional timber and masonry construction. The many internal walls that enclose the rooms participate in the lateral resisting system as **crosswall**s (see Section 2.3.2).

2.4 CONCLUSION: THE OBJECTIVE OF CURRENT SEISMIC DESIGN CODES

Much of the emphasis in this book has been on the behaviour of structural systems that are forced beyond their elastic limit. This is why **damping** and **ductility** are given such prominence.

Other than engineers with earthquake-specific training, few people are aware that current building codes in all earthquake-affected countries assume that a predictable level of earthquake shaking will cause structural damage in code-conforming new buildings. The objective of the building codes is to preserve life rather than prevent damage to property. Thus, it is building collapse that is mainly to be prevented. This is not true with wind forces, for example. Buildings must be designed to resist the full force of the largest winds expected at the site (other than tornados) without suffering structural damage.

Total avoidance of damage from "design-level earthquakes" in an earthquake area is an unsustainable objective because it is unaffordable and technically infeasible. The forces generated by earthquake vibrations are simply too great to require all ordinary code-conforming buildings to ride out the quake without damage. Thus, the earthquake forces used to calculate the code-conforming design are reduced from the full earthquake forces expected at any given site based on the ductility of the structural system.

Only nuclear power plants, and those other structures specifically designated to meet "immediate occupancy" standards, are designed so that their primary structural systems would remain within their elastic capacity when subjected to the forces from a design-level earthquake. All others are expected to deflect into the non-linear range, which means that damage will occur. "**Ductility** factors" have been included in the codes to represent the capacity of different materials and systems to behave in a ductile manner. Such factors have been derived for the modern structural materials of steel and reinforced concrete.

Unreinforced masonry naturally falls very low on this scale, but **ductility** factors to represent the better behaviour of *taq* and *dhajji dewari* when compared with unreinforced masonry are notably absent from building codes and engineering manuals. The good news is that in India, and more recently in Pakistan, at least for certain limited building types and sizes, modified versions of these two systems have been incorporated into provisions of the Building Codes and government guidelines. After the 2005 earthquake, both the Indian and Pakistan governments have produced manuals that include versions of both systems. However, in both countries this acceptance is limited to either one- or two-storey structures, and intended to be limited to construction in rural areas. As will be explained in Chapter 3, this is far more restrictive than it needs to be to take full advantage of these systems.

2.3.5b & 2.3.5c This new shopping centre in Srinagar shows many of the things not to do for earthquake safety. The walls are discontinuous to the ground, making for a complex load path through floors and walls. In addition, the concrete components, a detail of which is seen on the right, appear to be disconnected, and not part of a rational structural system. The masonry also most likely is not reinforced.

2.4.1 A building in Montorio nei Frentani, Italy, with "seatbelt strap" shoring installed after the 2002 Molise earthquake. This system illustrates how the timber bands in taq *construction work to hold the masonry walls together. The timber lacing in* taq *construction is significantly more effective than exterior straps because of its interlocked connection with the floor diaphragms.*

3.1a This barn in Uri shows damage to the unreinforced rubble stone wall below a dhajji wall that remained intact except for the loss of some panels. It is worth comparing this with the view of Kamsar near Muzaffarabad shown in Figure 3.1d. The stones in this wall are almost cubic, and thus vulnerable to the kind of damage seen there, but the stability of the dhajji construction, which remained bearing on the stone wall below, helped to keep the inside layer of the ground-floor stone wall, as well as the first-floor level from collapsing.

WHY IT WORKS: *TAQ* AND *DHAJJI DEWARI* IN
EARTHQUAKES

These mixed modes of construction are said to be better against earthquakes (which in this country occur with severity) than more solid masonry, which would crack.

Frederick Drew
The Jummoo and Kashmir Territories, 1875

3.1 THE 8 OCTOBER 2005 KASHMIR EARTHQUAKE

The Kashmir earthquake was one of the most destructive earthquakes in world history. The death toll from this magnitude 7.6 earthquake was approximately 80,000 and over 3 million were left homeless. In a region known to be so vulnerable to earthquakes, it is reasonable to ask: Why did both the masonry and reinforced concrete buildings in the area prove so vulnerable to collapse? Why did over 80,000 people lose their lives in what is a largely rural mountainous region? Why in Pakistan alone did 6,200 schools collapse onto the children at the time of morning roll call? This kind of scenario has played out repeatedly over recent decades in other earthquakes around the world, in cities and rural areas alike. Ironically, even as the knowledge of earthquake engineering has grown and become more sophisticated, earthquakes have taken an increasing toll in places where steel and reinforced concrete construction has displaced traditional construction.

After the 2005 earthquake, international teams of engineers and earthquake specialists fanned out over the damage districts on both sides of the Line of Control and returned with reports on the damage to different types of structures. Most of these reports focused on the Pakistan side of the Line of Control because the epicentre of the earthquake was northwest of Muzaffarabad. In that area, which has a high population density, the death and destruction was far more extensive than on the Indian side.

None of these reports covered timber-laced traditional construction of any type. The reason for this is superficially explained by the following exchange between Marjorie Greene of the Earthquake Engineering Research Institute (EERI), an international NGO, and various local officials and technical experts in Pakistan three months after the earthquake. She asked if they were aware of any examples of traditional timber-laced construction of any type in the earthquake-affected area. The officials answered that they were "unaware of any, but in years past, there may have been".

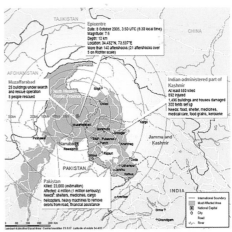

3.1b The 2005 earthquake-affected area. Source: Reliefweb Map Centre, Office for the Coordination of Humanitarian Affairs, United Nations, October, 2005.

3.1c Reinforced concrete building collapsed by the 2005 Kashmir earthquake in Balakot, North West Frontier Province, Pakistan. Photo by Dr. Robin Spence, UK.

3.1d Complete collapse of a cluster of modern limestone masonry and concrete houses in Bagh, Pakistan Administered Kashmir. The stones were shaped into blocks that were too short to build a properly bonded wall. Without timber lacing or some other system to confine the stones, this kind of masonry is extremely vulnerable. Photo by Maggie Stephenson, UN-HABITAT, 18 October 2005.

If the local experts were unaware of such construction, it is unlikely that it would be found by the foreign reconnaissance teams. The problem, however, goes deeper than that. It is actually less likely one of observation than it is one of orientation. Many would condemn traditional construction as unfit even had they seen examples of it still standing. Indeed, in the heart of the damage districts, *taq* and *dhajji dewari* construction was not as common as in Srinagar and the Vale of Kashmir, but it did exist. Its relative rarity may be a function of the fact that timber lacing was not needed on the rock and firm soil sites in the mountain regions to counteract differential settlement. In addition, Kashmiri timber-laced building traditions in the middle of the 20th century were displaced by standard unreinforced masonry construction, which in its turn has been more recently displaced by reinforced concrete.

Reinforced concrete frame construction now remains the only system that most local engineers are trained to design. Most buildings of 20th-century origin in both the Cantonment area of Uri, on the Indian side, and in Muzaffarabad and Balakot on the Pakistan side, for example, were of either unreinforced masonry or reinforced

3.1e An ironic example of where timber window frames proved stronger than solid masonry of this unreinforced stone masonry school, in Malot near Bagh, Pakistan Administered Kashmir, near the epicentre of the 2005 earthquake. Complete collapse in the 2005 earthquake was averted only because the wooden shutters of the windows were closed at the time of the earthquake. Photo by Maggie Stephenson, UN-HABITAT.

This, and many other rural schools were constructed with dressed stone facades, but with rubble stone interior layers and without attention to the need for bond stones and continuous ring beams either of timber or reinforced concrete.

3.1f The main mosque for the Uri Cantonment was less than 25 years old when it collapsed in the 2005 earthquake.

3.1g *The State Bank of India building in central Uri, of unreinforced rubble stone construction, was completely destroyed, whereas the* dhajji *wall of the barn immediately to the right of it, shown in Figure 3.1a, remained erect despite the collapse of the unreinforced rubble stone wall below.*

concrete, and many of these collapsed onto their occupants or were damaged beyond repair.

Despite the pervasive belief by the experts that timber-laced traditional construction did not exist, there are significant numbers of traditional buildings within the damage district. Most of these are hidden away within the mountain areas both in Pakistan and India. Moreover, in Pakistan, almost completely hidden from the view of the many engineers and government officials who had surveyed the district, an indigenous revival of traditional *dhajji* and *bhatar* construction has emerged after the earthquake – a revival that has grown out of the example of its good performance in the earthquake. This will be described in Sections 3.3 and 3.5, and also in Chapter 5.

In some ways, this lack of knowledge of the vernacular building systems in the earthquake area is not a surprise. It parallels a widespread lack of interest in such systems that exists in many countries which have recently experienced the rapid transformation from traditional materials and methods of construction to reinforced concrete. Of course, those buildings that exhibited good performance were often surrounded by many more houses of rubble stone and poor construction that did collapse, giving the impression of uniform devastation. It is also important to remember that *dhajji* walls resist earthquakes by deforming, which causes plaster and stucco to slough off in great quantities to the consternation of the occupants. To those lacking an understanding of this mechanism, the extent of this cosmetic damage gives the impression that the buildings are structurally unstable. In the case of both unreinforced masonry or reinforced concrete buildings, even less visible damage would commonly be an indicator of risk of collapse, but this is not so in the case of timber-laced traditional masonry construction. This has frequently led to unwarranted and unnecessary condemnation and evacuation, with the resulting high costs and unmet needs for humanitarian relief.

All one has to do is read the news accounts of the disaster to see how collapsed rubble masonry dwellings represented what was thought to be all of the rural building stock. The few surviving structures were ignored by surveyors intent only on photographing earthquake damage. The fact that significant numbers of the rural vernacular buildings did not fall down was not seen as sufficient reason to explore why they survived when so many others around them did not. There are always some winners and losers in an earthquake. Rarely is every building lying on the ground, so the survey teams inevitably concluded that the still standing ones were probably on the verge of collapse anyway, especially when their plaster had fallen off.

However, had these same inspectors looked more closely, they would have discovered that the difference between the winners and the losers was often not just chance, but rather the existence of timber lacing – either timber runners laid into rubble stone bearing walls, or lightweight timber frames nogged with a mixture of rubble stone and mud. In Pakistan, where the damage was the greatest, these early systems were not noticed by the government engineers and public officials because they represented a building tradition that was no longer accepted and that most engineers at the governmental and donor funding level did not want to endorse because there was no engineering data on it. This was also consistent with standard government policy because the further cutting of wood in the mostly deforested country was not allowed, and reinforced concrete had been embraced as the safe and modern way to build and had become the almost exclusive focus of engineering training in the universities.

As a consequence, after the earthquake, the Government of Pakistan began to withhold reconstruction assistance funds from those people who proceeded to rebuild with *dhajji* or other timber-laced systems. For over a year after the earthquake, only those who followed the government's approved plans were allowed to obtain government assistance. This belief in the efficacy of reinforced concrete and concrete block continued despite its abysmal performance in Muzaffarabad, Balakot, and over the whole region in the earthquake. (Pakistan government surveys show that in Muzaffarabad 89% of residential structures were destroyed. Nearby, in Balakot, the damage reached 100%. Throughout Pakistan Administered Kashmir, 84% of the housing units were reported as damaged or destroyed.) For construction in the rural areas of Kashmir, as will be explained below and in Chapter 5, the Government of Pakistan eventually changed their policy to include and even encourage the use first of *dhajji*, and later of *bhatar* construction as "compliant" systems eligible for financial assistance.

3.2 The Earthquake Resistance of *Taq* Construction

Returning to the Indian side of Kashmir, one of the most important of the post-earthquake reconnaissance reports was published by EERI. This report was written by Professors Durgesh C. Rai and C. V. R. Murty of the Indian Institute of Technology, Kanpur and published in December 2005 as part of EERI's *Learning from Earthquakes* report on the Kashmir earthquake. The quotations below from the authors were based on observations made during the first several weeks after the earthquake. Describing *taq* construction, which they observed in the damage district on the Indian side of the Line of Control, Professors Rai and Murty observed:

> *In older construction, [a] form of timber-laced masonry, known as* Taq *has been practised. In this construction large pieces of wood are used as horizontal runners embedded in the heavy masonry walls, adding to the lateral load-resisting ability of the structure...Masonry laced with timber performed satisfactorily as expected, as it arrests destructive cracking, evenly distributes the deformation which adds to the energy dissipation capacity of the system, without jeopardizing its structural integrity and vertical load-carrying capacity.* (Rai and Murty, 2005)

It is interesting to compare their observation with that of Professors N. Gosain and A.S. Arya after an inspection of the damage from the Anantnag Earthquake of 20 February 1967, where they found buildings of similar construction to Kashmiri *taq*:

> *The timber runners...tie the short wall to the long wall and also bind the pier and the infill to some extent. Perhaps the greatest advantage gained from such runners is that they impart ductility to an otherwise very brittle structure. An increase in ductility augments the energy absorbing capacity of the structure, thereby increasing its chances of survival during the course of an earthquake shock.* (Gosain and Arya, 1967)

These two reports are separated by almost 40 years. Gosain and Arya ascribe a kind of ductile behaviour to the timber-laced masonry and even say that the timbers "impart ductility" and augment "energy absorbing capacity", while Rai and Murty use the term "energy dissipation capacity" to describe the same phenomenon. The different ways of describing this behaviour simply reflect changes in terminology, as the word "ductility" is more scientifically correct when used to describe an attribute of a single material rather than that of a combination of materials; but the basic phenomenon remains the same, and other noted scholars have made similar observations in other countries. In Turkey in 1981, Professor Alkut Aytun credited the bond beams in that country with "incorporating ductility to the adobe walls, substantially increasing their earthquake resistant qualities", and illustrated his point with a

3.2a Photo by Turkish Professor Alkut Aytun showing a two-storey brick house with timber bond beams spanning the ground rupture from the 1970 Gediz earthquake in western Turkey. In the caption to this image, he observes that this shows the effectiveness of timber bond beams for both fired or unfired brick masonry (Aytun, 1981).

3.2b The side of a timber-laced masonry building near Uri after the 2005 earthquake: The quake caused the cracking of the modern rigid cement plaster from the otherwise non-destructive shifting of the underlying timber-tied masonry.

photo of a house that spanned the fault rupture in the 1970 Gediz earthquake, where the timber lacing held the masonry bearing wall together despite approximately 50cm of vertical ground displacement (Aytun, 1981) (see Section 2.2.4).

The concept of ascribing ductility to a system composed of a brittle material – masonry – is a difficult one for many modern engineers. While a steel coat hanger is ductile, which can be seen when it is bent beyond its elastic limit, a ceramic dinner plate is brittle. So how can masonry, which on its own is inarguably made up of brittle materials, as shown in Figure 2.2.4a, be ductile? Rai and Murty in 2005 avoided the use of the term "ductile" probably because the materials in *taq* are not ductile and do not manifest plastic behaviour. However, what makes timber-laced masonry work well in earthquakes is its ductile-like behaviour as a system. This behaviour results from the energy dissipation because of the friction between the masonry and the timbers and between the masonry units themselves.

This friction is only possible when the mortar used in the masonry is of low-strength mud or lime, rather than the high-strength cement-based mortar that is now considered by most engineers to be mandatory for construction in earthquake areas. Strong cement-based mortars force the cracks to pass through the bricks themselves, resulting in substantially less frictional damping and also rapidly leading to the collapse of the masonry. Arya made this difference clear when he said: "Internal damping may be in the order of 20%, compared to 4% in uncracked modern masonry (brick with Portland cement mortar) and 6%-7% after the masonry has cracked." His explanation for this is that "there are many more planes of cracking…compared to the modern masonry" (see Section 2.2.3).

The difference between cement mortar and weaker traditional mortars made of mud or hydrated lime has been a significant issue in the field of building conservation for many years, but the debate over the efficacy of the weaker mortars has been different in areas where earthquakes are a risk. As can be seen in this quotation concerning premodern masonry construction in the United States, there are attributes to mortar other than compressive and tensile strength that are even more fundamental to the long-term stability and conservation of the masonry:

> It is essential to distinguish between hard and soft mortars. The use of lime-sand mortar … was soft enough to furnish a plastic cushion that allowed bricks or stones some movement relative to each other… A cushion of soft mortar furnished sufficient flexibility to compensate for uneven settlement of foundations, walls, piers and arches: gradual adjustment over a period of months or years was possible. In a structure that lacks flexibility, stones and bricks break, mortar joints open and serious damage results. (Harley McKee, 1973)

In areas subject to earthquakes, engineers have often sought to specify strong cement-based mortar. However, in the larger earthquakes, the strength of the mortar ceases to be helpful once the walls begin cracking, as they inevitably do in a strong earthquake. It is then that the "plastic cushion" and other attributes described by Harley McKee become more important. Perhaps most important is that the masonry units – the stones or bricks – be stronger than the mortar so that the onset of shifting and cracking is through the mortar joints, and not through the bricks. Only then can the wall shift in response to the earthquake's overwhelming forces without losing its integrity and vertical bearing capacity. With timber-laced masonry, it is important to understand that the mortar is not designed to hold the bricks together, but rather to hold them apart. It is the timbers that tie them all together. The benefits of energy dissipation are gained from the non-destructive friction and cracking that can take place in a masonry wall that is surrounded and thus confined by the timber bands.

Professors Rai and Murty also point out that the timber lacing "arrests destructive cracking" in the masonry walls. This phenomenon may have been understood as far back as in ancient Rome, if not earlier, where concrete walls were constructed with horizontal bands of brick that extended through the walls subdividing the concrete. Crack-stopping masonry construction technology can also be found in medieval construction in Istanbul, with similar bands of brick in the Theodesian walls which still exist around the oldest part of the city. Each of these bands is called a *hatıl*, (plural *hatıllar*) the Turkish term still used to identify the horizontal timber ring beams in masonry that were commonly used throughout the Ottoman Empire, the cultural influence of which may be in part responsible for the *cator*, *bhatar*, and *taq* construction found in Pakistan and India. (For illustrations of the Turkish examples, see Langenbach, 2007a.)

3.3 *TAQ* TIMBER LACING IN CODES AND GUIDELINES

3.3.1 INDIA: Although not identified by their Kashmiri names, both the *taq* and *dhajji dewari* systems have been recognized and adopted into several of the Indian Standard National Building Codes. Dr Anand S. Arya, a principle author of the Indian building codes, undertook research at IIT-Roorkee on the construction systems and was influential in their inclusion into the Indian Standard code. He has reported that "timber runners" (the horizontal timbers laid into masonry walls like those in *taq*) were introduced into the code for non-engineered construction in 1967.

In 1980, timber ring beams of the same kind of ladder-like configuration as are illustrated in the Indian Standard codes were described and illustrated in the internationally recognized document

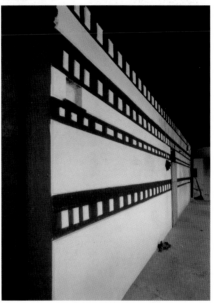

3.2c & d In Ahmedabad, this timber lacing in the 17th-century "Old Haveli" of the Dwarka Dheesh Mandir is treated as a part of the architectural ornament. This and many other timber-laced bearing wall and timber frame with infill structures in Ahmedabad did well in the 2001 earthquake, with no collapses of occupied structures and no deaths. By contrast, many nearby reinforced concrete buildings, almost all of which had been constructed during the most recent decade, collapsed.

3.2e While most taq *buildings in Srinagar are constructed of fired brick, the lower storeys of this large house are of stone and do not observe the modular window bay layout from which the term* taq *is derived. Timber-laced stone construction is more common in Baramulla and the surrounding rural mountain areas further to the west.*

published by the International Association of Earthquake Engineering based in Japan. This document was revised in 1986 by an international committee composed of Teddy Boen from Indonesia, Yuji Ishiyama from Japan, A. I. Martemianov from the USSR, Roberto Meli from Mexico, Charles Scawthorn from the USA, Julio N. Vargas from Peru and Ye Yaoxian from China, and re-published under the title *Guidelines for Earthquake Resistant Non-Engineered Construction*. The committee was chaired by Professor Anand Arya from IIT-Roorkee, India. (Arya et al, 1986). (The National Information Center of Earthquake Engineering (NICEE) in Kanpur has now made it available on the web in Hindi and in English.) The same concepts also have been recognized in the Turkish codes as early as 1972 (Aytun, 1981), and in the Nepal Building Code, which was drafted in 1994.

The most recent Indian codes date from 1993. A modified form of *taq* construction can, as mentioned above, be found in the Indian Standard Building Codes *IS 13827: Improving Earthquake Resistance of Earthen Buildings – Guidelines*, and *IS 13828: Improving Earthquake Resistance of Low-Strength Masonry Buildings – Guidelines*. Low-strength masonry "includes fired brick work laid in clay mud mortar and random rubble; uncoursed, undressed or semi-dressed stone masonry in weak mortars; such as cement-sand, lime-sand and clay mud". (The inclusion of the word "guidelines" in the titles of these codes is emblematic of the fact that for the most part, building codes in India remain advisory documents, rather than mandatory rules, and also a recognition of the fact that these particular codes address types of rural construction for which it would be difficult to enforce mandatory rules in any case.)

After the 2005 Kashmir earthquake, the Government of India, Ministry of Home Affairs gave further recognition of variants of the traditional Kashmiri systems by including them in their publication that followed the earthquake, *Guidelines for Earthquake Resistant Reconstruction and New Construction of Masonry Buildings in Jammu and Kashmir State* (Arya and Agarwal, 2005). The two 1993 codes and the 2005 post-earthquake manual all specify that seismic bands (also known in English as tie beams, ring beams or bond beams) are recommended for use in new masonry construction. The timber versions of these bands shown on page 14 of the 2005 *Guidelines*, and in *IS 13828*, Section 8 are very similar to what has existed in *taq*, *bhatar*, and *hatil* timber-laced construction for centuries.

IS 13827 for earthen buildings, Section 11.1 specifies that "two horizontal continuous reinforcing and binding beams or bands should be placed, one coinciding with lintels of door and window openings, and the other just below the roof in all walls in seismic zones III, IV, and V." This code specifies that these bands be made of timber. For thin walls, a single timber may be used with diagonal corner braces, but for thicker walls, two timbers are to be laid on either side of the wall and connected with short pieces. In *IS 13828* for low-strength masonry, the use of timbers is specified as an alternative to the use of

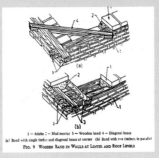

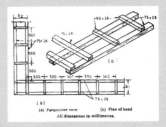

(1) Illustrations from IS 13827 showing timber lacing recommended for earthen buildings.

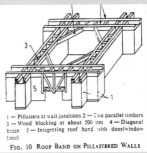

(2) Illustration from IS 13828 showing timber lacing recommended for low-strength masonry buildings.

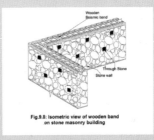

(3) Illustration from 2005 Guidelines for Earthquake Resistant...Masonry Buildings in... Kashmir showing recommended use of timber bands in stone masonry buildings.

3.3b Traditional bhatar *construction in the Swatt Valley town of Behrain used to stabilize a rounded random rubble stone wall construction which otherwise is extremely at risk in earthquakes.*

*3.3c Early 20th-century fort in Besham, Pakistan after the 2005 earthquake, showing how the collapse of this dry-laid rubble stone wall was prevented by the timber lacing. This demonstrates the effectiveness of the timber tie beams (*bhatari*) in holding the earthquake-damaged rubble stone wall together.*

3.3d Training of craftsmen and owner-builders in a rural area of Pakistan Administered Kashmir. Photo by SDC.

a continuous band of reinforced concrete. The placement of the seismic bands is the same as in *IS 13827*.

The sizes of the timbers for the reinforcing bands specified in both codes, which range from 2" x 3" (50 x 75mm) to 3" x 6" (75 x 150 mm), are smaller than those used historically. For the *taq* houses, the sizes were probably dictated by the difficulty of cutting timber into boards before the arrival of power saws, and the need for timber thickness for the scarf joints that pre-date the use of metal straps. Today, connections can be accomplished with steel hardware and nails, and the timber can be efficiently sawn into boards at a sawmill. This may be the rationale for the smaller dimensions. (It should be noted that conformity with the smaller dimension makes the notching together of the timbers difficult to accomplish, and relying on nails which eventually rust may not be adequate if an earthquake should occur years later.)

3.3.2 PAKISTAN: The recognition and adoption of timber-laced masonry construction has followed a different and more recent trajectory in Pakistan where the worst effects of the 2005 Kashmir earthquake resulted in a humanitarian crisis that required creative solutions to the problem of housing construction in remote rural areas. The Government of Pakistan reported that approximately 75,000 people were killed and 70,000 injured in Northern Pakistan, including Pakistan Administered Kashmir, and 600,000 houses and 6,298 schools were destroyed (ERRA, 2006). On the Indian side of the Line of Control, 1,350 were killed and 6,300 injured.

Immediately after the earthquake, the Government of Pakistan established the Earthquake Reconstruction and Rehabilitation Authority (ERRA) to "facilitate the rebuilding and repair of damaged infrastructure, including housing, roads, bridges, government buildings, schools and hospitals". (ERRA, 2007) There were and continue to be two important mantras that guide the activities of this organization, as well as the associated international organizations that have come to provide aid in the earthquake-stricken areas of Pakistan. The first is "build back better" and the second is that the housing rebuilding process would be "owner-driven".

Although not always achieved, the "build back better" goal is a fundamental goal of disaster rebuilding efforts in general, but the "owner-driven" concept has turned out to be of crucial importance. What it means is that the objective was not for the government or NGOs to come in and provide the rebuilt housing, but rather only to provide training and funding for the owners themselves to rebuild their own housing on their own properties. The government retained responsibility for the inspection to allow only approved earthquake resistant construction to receive financial assistance (ERRA, 2007 and www.erra.gov.pk). This may not seem noteworthy except that throughout the world there are many examples of enclaves of culturally insensitive identical houses built after earthquakes where the houses are lined up in military-like rows, such as those described in Section 4.4 (see also Langenbach and Dusi, 2006c). ERRA wisely resisted pressure by some donors and NGOs which wanted to undertake the construction directly. In 2007, General Nadeem Ahmed, former Deputy Chairman ERRA, who supported and maintained the owner driven policy and ensured it was responsive to field needs and community priorities, was awarded the UN-HABITAT scroll of honour. UN-HABITAT has reported that the "owner-driven" practice in Pakistan has resulted in a massive broad base of new skills and knowledge, as well as the relearning of forgotten crafts.

In Pakistan, a major contribution in support of ERRA was made by teams fielded by the Pakistan Poverty Alleviation Fund and the National Rural Support Programme (NRSP). The international organizations include UN-HABITAT, the Swiss Agency for International Development and Cooperation (SDC), the National Society for Earthquake Technology, Nepal (NSET), Architecture & Développement, France (A&D), the French and Belgian Red Cross, and over 60 other NGOs. These organizations provided major assistance in setting up training centres in the damage district and carrying out the trainings of carpenters and masons as well as owner-builders (see Figures 3.3d, 3.3e and 3.3f). In the case of UN-HABITAT, a major portion of the staff contribution was made by Kashmiris, including Sheikh Ahsan Ahmed, S. Habib Mughal and Hamid Mumtaz, who supported the approval of traditional construction.

Government endorsement of traditional techniques was not immediate, however. In May of 2006, only seven months after the earthquake, ERRA published a manual designed to establish a government-approved standard for earthquake-resistant construction that would be considered "compliant" and thus eligible for financial reimbursement from the government. This document, *Guidelines For Earthquake Resistant Construction Of Non-Engineered Rural And Suburban Masonry Houses In Cement Sand Mortar In Earthquake Affected Areas*, specified the type of reinforced masonry that would be eligible for government assistance.

3.3e, f & g Views of a training workshop on bhatar *construction for local masons in Battagram, Pakistan, in spring and summer 2007, prepared by the French Red Cross with Architectes et Développement. Photos by the French Red Cross.*

As the post-disaster rebuilding process proceeded, the rigid adherence to the one construction typology presented increasing problems as the relief efforts moved from the urban areas into increasingly remote rural areas. As was reported in the ERRA 2007 Environmental Assessment (ERRA, 2007), "Of the total housing stock, 84 percent was damaged or destroyed in Pakistan Administered Kashmir and 36 percent was damaged or destroyed in NWFP [North

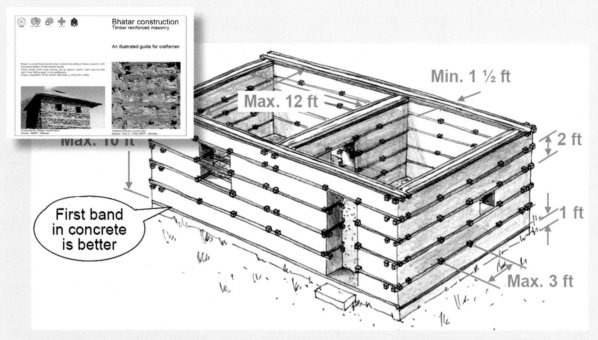

3.3h Drawing from the training manual Bhatar Construction, an Illustrated Guide for Craftsmen. *This manual was created by Swiss Architect Tom Schacher of SDC and published by the Government of Pakistan for use in the training of rural craftsmen for the reconstruction of earthquake-damaged homes. Source: Tom Schacher and ERRA, Government of Pakistan.*

West Frontier Province]. The affected houses were predominantly rural, with urban units accounting for only 10 percent of the total. Much of the rural housing was located on steep slopes difficult to access." A requirement that government assistance be limited to reinforced masonry with cement mortar would mean that the materials would have to be transported deep into the countryside where sometimes there were no roads to meet this requirement (see Figure 3.3k). As a result, many of the families rendered homeless began rebuilding destroyed rubble masonry houses using traditional timber-laced masonry.

At this stage, individuals from the NGOs with their Pakistani colleagues had a chance to examine the various forms of traditional construction that had done well in the earthquake and which were already being copied by the survivors for new construction to replace their destroyed rubble stone homes. Together, they began to work with ERRA and NESPAK to establish ways to broaden the compliant construction types, first to accept *dhajji* construction and then later to accept *bhatar*. One of the difficulties, as reported by UN-HABITAT, was the fact that *dhajji* and and *bhatar* construction had not been the subject of engineering research and no generally accepted analytical tools had been developed for it. In an effort to provide some professional credibility for the systems, UN-HABITAT prepared the report shown in Figure 3.3l in which their engineering consultants state, "Because of the inherent variability and complexity of the individual materials...it is not possible to accurately calculate or model the structural behaviour of the *bhatar* system. As the *bhatar* system relies on structural stability and energy dissipation rather than strength characteristics, standard calculation techniques appropriate for dynamic analysis of engineered structures have limited validity when applied to *bhatar* construction" (see sidebar on page 50 for more excerpts). This explains why many engineers, including those representing the World Bank and other donor agencies have difficulty accepting traditional construction.

With the support of Army personnel who could see the practicality of using the local vernacular in the mountain areas and the urgency of the need for disaster assistance funding in this historically volatile region, *bhatar* was approved by ERRA in July 2007. While overseeing a programme covering the construction of 630,000 new and repaired houses, Waqas Hanif, the ERRA Programme Manager for Rural Housing, came to embrace both *dhajji*

and *bhatar* and thus was key in ensuring both were approved as compliant. Despite the effort that it took, architect Tom Schacher observed that "the readiness of the engineering consultants to the government to review their dogmas and approve construction practices hitherto unknown to them and for which they often didn't have the required scientific evidence was extraordinary." (Schacher, 2008)

This history in Pakistan is significant for a number of reasons. Not only has the earthquake made it clear to the government and affected citizens alike that there is a need for more structurally sound and earthquake resistant construction, even in rural areas, it has also served to bring urbanized and university-educated architects and engineers into contact with the culture and indigenous building crafts characteristic of the rural regions. Most had long identified non-engineered traditional masonry construction of all types as archaic and unsuitable for contemporary living, particularly in an earthquake area, but after such a devastating earthquake, they could witness for themselves what survived and what failed. The interaction between the foreign humanitarian technical support teams both with the local engineers and government officials and the local population was crucial in what actually became a creative two-way technology transfer. Before either *dhajji* or *bhatar* could be adopted, both the foreign and the Pakistani professionals had to jettison their pre-existing prejudices to accept and improve upon premodern systems that were taught to them by the local people themselves. This stands as a remarkable example of openness, creativity, and acceptance at all levels.

3.3i & j Completed building with bhatar *following the Pakistani government-endorsed guidelines constructed by SDC as a demonstration of approved* bhatar *detailing. Photos by Tom Schacher, SDC.*

3.3k Cement blocks for housing reconstruction being moved by footpath to a remote site. Source: ERRA, 2007.

3.3l This report was produced by UN-HABITAT for the Pakistan Government.

Engineering Research at IIT-Roorkee

3.3m & n Full-size models of rubble stone low-strength masonry houses, 3.3m with seismic bands, and 3.3n the same construction but without seismic bands. Both were tested on the IIT-Roorkee shock table. Source: Agarwal et al, 2002.

A Structural Engineering Review of why *Bhatar* Performs Well in Earthquakes

Excerpts from: *Build Back Better – Bhatar: Background and Rationale*, by, Dominic Dowling PhD, Engineer, Pierre-Yves Pere, Architect, French Red Cross, and Pierre Perrault, Engineer, Belgian Red Cross, for UN-HABITAT, June 2007.

4.1 Structural Components: The *bhatar* system consists of stone masonry walls reinforced with horizontal timber ladder-bams, which combine to resist and dissipate the energy induced during an earthquake....

4.2 Structural Characteristics: The combination of dry-stacked stone masonry and timber *bhatar* bands provides strong resistance to seismic forces. Key structural characteristics include:

> (1) Resistance to out-of-plane bending is provided by the horizontal timber elements…embedded in the wall…a similar structural function as reinforced concrete beams/bands in reinforced or confined masonry. (2) Resistance to out-of-plane…**overturning**..is provided by the inherent stability of the walls with a height to thickness ratio commonly ranging from 4 to 6. (3) Resistance to …shear forces is provided by the inherently high friction in the system. (4) Resistance to vertical corner cracking (the most common and critical damage pattern in unreinforced masonry walls) provided by the interconnected *bhatar* bands at the corners. (5) Resistance to delamination of the masonry walls…is provided by the *bhatar* bands... (6) Resistance to cyclic loading is provided by the timber elements which possess excellent tensile and elastic characteristics, and the friction within the stone masonry.

The fundamental principle of the *bhatar* system is dissipation of energy through friction (shear). Physical bonds exist between the different components, so the shear capacity of the structure is proportional to the coefficient of friction between adjoining elements (stone-stone and stone-timber), the cross-sectional area of the wall, and the gravitational load. For *bhatar* structures, all of these factors are inherently present (dry-stacked stone masonry, interlocking timber elements, and wide and heavy walls) so the capacity for dissipation of energy through friction is very high.[10]

4.3 Structural Calculations: Because of the inherent variability and complexity of the individual materials, component interactions and forms of construction, it is not possible to accurately calculate or model the structural behaviour of the *bhatar* system. As the *bhatar* system relies on structural stability and energy dissipation rather than strength characteristics, standard calculation techniques appropriate for dynamic analysis of engineered structures have limited validity when applied to *bhatar* construction. Of greater value is the vast amount of empirical data available from post-earthquake reconnaissance, and historic evidence.

3.4a Earthquake-damaged building in Baramulla in 2005 showing collapse of unreinforced masonry leaving the dhajji *reinforced wall suspended above. The lower part of the walls lacked ties securing the floors to the walls, thus the building was missing a critically important requirement for protection against collapse in earthquakes.*

3.4 THE EARTHQUAKE RESISTANCE OF *DHAJJI DEWARI* CONSTRUCTION

In the 2005 EERI report, Professors Rai and Murty commented more extensively on how *dhajji dewari* construction was affected by the earthquake:

> *In Kashmir traditional timber-brick masonry construction consists of burnt clay bricks filling in a framework of timber to create a patchwork of masonry, which is confined in small panels by the surrounding timber elements. This timber lacing of masonry, which is locally referred as dhajji-dewari has excellent earthquake-resistant features. The resulting masonry is quite different from typical brick masonry and its performance in this earthquake has once again been shown to be superior with no or very little damage. No collapse was observed for such masonry even in the areas of higher shaking.*

They go on to explain the reason for this good behaviour:

> *The presence of timber studs, which subdivides the infill, arrests the loss of the portion or all of several masonry panels and resists progressive destruction of the rest of the wall. Moreover, the closely spaced studs prevent propagation of diagonal shear cracks within any single panel, and reduce the possibility of out-of-plane failure of masonry of thin half-brick walls even in the higher storeys and the gable portion of the walls (Rai and Murty 2005).*

3.4b Adapazari, Turkey after the 1999 Kocaeli earthquake showing surviving traditional structure next to pancake-collapsed modern reinforced concrete blocks. The collapsed buildings extend beyond the mosque visible way in the distance behind the tree. (The mosque and its minaret are unreinforced masonry.)

3.4c Interior dhajji *wall in a house 13 miles from Uri, in the heart of damage district of the 2005 earthquake, showing the "working" of the timber frame against the masonry panels causing non-destructive cracking of the overlying plaster.*

Dhajji dewari is timber frame construction rather than masonry bearing wall construction. Thus the vertical loads are transferred to the ground primarily, but not exclusively, through the frame. However, the masonry does form an integral part of the structural system, sharing the vertical load path with the timber frame. As has already been explained, this infill masonry serves a primary role in the case of lateral earthquake loads.

In the damage district of Kashmir on both sides of the Line of Control, there were enough buildings of *dhajji* construction to observe the effects of the earthquake on the construction system. In the mountains on the Pakistan side, the infill material is more commonly rubble stone set in mud mortar, while on the Indian side, particularly in the Vale of Kashmir where clay is abundant, the use of fired and unfired clay brick is more common.

In Turkey, *hımış*, which is similar to *dhajji*, survived in the heart of the damage districts of the two large earthquakes around the Sea of Marmara in 1999 when hundreds of surrounding concrete buildings collapsed, killing over 17,000 people (the unofficial death toll was much higher). This finding was confirmed by Turkish researchers who conducted a detailed statistical study in several areas of the damage district. They found and documented a wide difference in the percentage of modern reinforced concrete buildings that collapsed, compared to those of traditional construction.

In one district in the hills above Gölcük where 60 of the 814 reinforced-concrete, four-to-seven-storey structures collapsed or were heavily damaged, only four of the 789 two- or three-storey traditional structures collapsed or had been heavily damaged. The reinforced-concrete buildings accounted for 287 deaths, against only three in the traditional structures.

In the heart of the damage district in Adapazari, where the soil was poorer, research showed that 257 of the 930 reinforced concrete structures collapsed or were heavily damaged, and 558 were moderately damaged. By comparison, none of the 400 traditional structures collapsed or were heavily damaged, and 95 were moderately damaged (Gülhan and Güney, 2000).

Thus, both the Kashmir and the Turkey earthquakes provided a chance to examine how this type of light frame with masonry infill construction actually performed where the earthquake shaking was large enough to cause widespread destruction of other building types, from unreinforced masonry to reinforced concrete. Inspections of the interiors of the *dhajji* houses in both India and Pakistan, as well as in Turkey, provide a more complete understanding of the behaviour of *dhajji* as a structural system under earthquake loads. It was evident that the infill masonry walls responded to the stress of the earthquake by "working" along the joints between the infilling and the timber frame; the straining and sliding of the masonry and timbers dissipated a significant amount of the energy of the earthquake. The only visible manifestation of this internal movement was the presence of cracks in the interior plaster along the walls and at the corners of the rooms, revealing the pattern of the timbers embedded in the masonry underneath.

On the exterior, where there was usually no plaster coating, the movement of the panels often was not very visible. The movement was primarily along the interface between the timbers and the brick panels where a construction joint already exists. Because of the timber studs that subdivide the infill, the loss of masonry panels did not lead progressively to the destruction of the rest of the wall. The closely spaced studs prevented propagation of "X" cracks and reduced the possibility of the masonry falling out of the frame. Where it was observed that large sections did fall out, it could most often be attributed to rotted timbers or oversized panels or both, and the structures involved were often barns rather than houses.

An important additional factor in the good performance of the walls was the historical use of weak rather than strong mortar. For the same reasons as explained above for *taq* construction, the mud or weak lime mortar encourages sliding along the bed joints instead of cracking through the bricks when the masonry panels deform. This sliding also serves to dissipate energy and reduce the incompatibility between rigid masonry panels and the flexible timber frame.

The basic principle in this weak and flexible frame with masonry infill construction is that there are no strong and stiff elements to attract the full lateral force of the earthquake. The buildings thus survive the earthquake by not fully engaging with it. This "working" during an earthquake can continue for a long period before the degradation advances to a destructive level.

Thus the engineering principle behind the earthquake-performance of the *dhajji* walls is a simple one. The subdivision of the walls into many smaller panels with studs and horizontal members, combined with the use of low-strength mortar, prevents the formation of large cracks that can lead to the collapse of the entire infill wall, while the redundancy provided by the many interior and exterior walls that exist in a standard residential building reduces the likelihood of catastrophic failure of the frame.

3.4d Two-storey block of shops of dhajji *construction near Bagh on road to Thub, Pakistan Administered Kashmir, photographed after the 2005 earthquake. This building suffered only minor damage even though it was close to the epicentre.*

3.4e This reinforced concrete building, also a block of shops, was located near to the dhajji *building in 3.4d. It had been three storeys until the ground-floor collapsed in the 2005 earthquake. It was the only concrete building in the rural village of Thub, Pakistan Administered Kashmir.*

While these structures lack lateral strength, they possess lateral capacity. They respond to seismic forces by swaying, rather than by attempting to resist them with rigid components and connections. This is not an elastic response, but a plastic one. When these structures lean in an earthquake, they do so with incremental low-level cracking which is distributed throughout the wall by the interaction of the timber structural components with the confined masonry. In other words, although the masonry and mortar is brittle, the system – rather than the materials that make up that system – behaves as if it were "ductile".

Ordinary unreinforced masonry walls under lateral earthquake loads develop diagonal tension "X" cracks, which then can rapidly progress until the wall collapses. With *dhajji dewari*, the frame of studs and cross-pieces prevents the development of the large diagonal cracks, thus helping to hold the wall together against both in-plane and out-of-plane shaking.

In addition, the non-destructive movement of the panels along the construction joints between the panels and the frame introduces a large amount of friction, which in turn reduces the excitation of the building from the earthquake (see Section 2.2.3). This reduces the damage to the rest of the superstructure.

In earthquakes in Turkey, Greece and the former Yugoslavia, as well as the 2005 earthquake in Kashmir, timber frame and masonry infill construction has demonstrated a high degree of frictional damping over many cycles with relatively slow rates of degradation. *Dhajji dewari* thus avoids the drawbacks of bearing wall unreinforced masonry, which is heavy and subject to brittle failure in earthquakes and fracture from differential settlement.[11]

While in Kashmir it is difficult to isolate the influence that earthquakes may have had on the evolution of either the *taq* or the *dhajji dewari* technology, there are two examples of *dhajji*-like timber frame infill systems that indisputably were developed and promulgated (and even patented) specifically for their earthquake resistance. One is the *gaiola* frame in Portugal, otherwise now known as "Pombalino walls" after the Marquis de Pombal, who was responsible for its development after the devastating 1755 Lisbon earthquake (see sidebar). The other is the Casa Baraccata which was invented under similar circumstances after the Calabria earthquake of 1783 and later patented (Barucci, 1990).

3.4f This old commercial building in central Srinagar is constructed with two storeys in taq *and the top storey in* dhajji dewari. *This is a good combination because the lighter-weight construction at the top is tied together by the timber frame, while the timber-laced bearing walls of the* taq *construction benefit from the overburden weight of the* dhajji *construction above.*

An Eighteenth-Century "Invention" in Portugal

In eighteenth-century Portugal, the Marquis de Pombal builds *gaiolas* ("cages") to protect against earthquakes

The story of the *gaiola* in Portugal is an interesting one. After the 1755 Lisbon earthquake, Chief Minister Sabastião José de Carvalho e Melo, later the Marquis of Pombal, gathered a group of military engineers led by Manuel de Maia to determine the best manner of earthquake-resistant construction to use for the rebuilding. For this, they developed the *gaiola* ("cage"), which has since become known as Pombalino construction (Penn et al, 1995). The *gaiola* essentially is a well-braced variation of *dhajji dewari* construction (also see Section 1.3 & Figures 1.3n, o & r.)

After building and testing a prototype, the Marquis made its incorporation into the reconstructed buildings a requirement. The inspiration to use this system came most likely from the observation of surviving medieval masonry infilled timber-frame structures after the earthquake. This is consistent with the eyewitness report by Reverend Charles Davy, who alluded to the distinction between what stood and what fell in his comment: "With regard to the buildings, it was observed that the solidest in general fell the first" (Tappan 1914), a comment not unlike Arthur Neve's observations about Srinagar after the 1885 earthquake quoted earlier. Many of the new buildings with the *gaiola* were five or six storeys in height and most of these remain standing today, 225 years later. The significance of the Pombalino system lies in the fact that it was deliberately developed and selected as earthquake-resistant construction for a major multi-storey urban area.

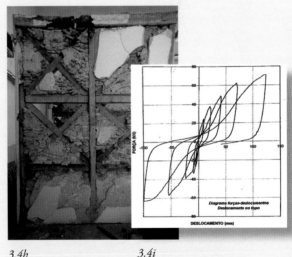

3.4g   3.4h   3.4i

In 1997, the seismic capacity of the *gaiola* was tested in the Portuguese Government's lab by subjecting actual wall sections to cyclical tests, one of which is in Figure 3.4h seen after the test. It was removed from a building in the Baixa district shown in Figure 3.4g, where all buildings were required to be reconstructed with *gaiola* walls. The test wall remained largely intact despite having been pushed cyclically beyond what would be expected from an earthquake. The almost complete loss of plaster shows that the forces were distributed across the wall section. The wide hysteresis loops (3.4i) are a measure of a large amount of friction-induced energy dissipation (hysteretic damping), and the increased height of each successive loop shows that the wall had not exceeded its maximum strength at the conclusion of the test (Cóias e Silva, 2002 & Santos 1997). (See Figure 2.2.3a and Section 2.2.3 for an explanation of hysteretic damping and the significance of loops of different shapes). In 2003, a research project at the Istituto per la Valorizzazione del Legno e delle Specie Arboree (IVALSA) in Italy involving a similar quasi-static cyclic tests on both brick and stone infilled timber wall specimens showed similarly robust results (Ceccotti, 2005).

3.5a Illustration from Ministry of Home Affairs Guidelines for Earthquake Resistant Reconstruction and New Construction of Masonry Buildings in Jammu & Kashmir State, *by A.S. Arya.*

An earlier version of this illustration can be found in the 1986 publication Guidelines for Earthquake Resistant Non-Engineered Construction, *produced by the International Association for Earthquake Engineering, based in Japan. This 1986 volume was the work of an international ad-hoc committee chaired by Anand S. Arya Professor of Engineering at IIT-Roorkee, India (see Section 3.3.3 for more information).*

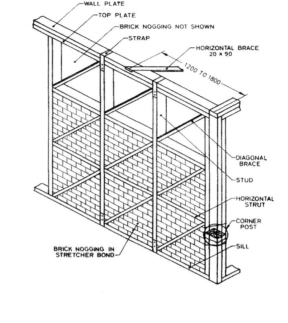

3.5b & c Photographs of construction training workshops given under the auspices of ERRA and UN-HABITAT. On the blackboard in 3.5b the different patterns of the timbers in dhajji *construction found in rural Kashmir can be seen. The model in 3.5c is used to teach timber joinery used for* dhajji *construction. Photos provided by UN-HABITAT.*

3.5 *DHAJJI DEWARI* IN CODES AND GUIDELINES IN INDIA AND PAKISTAN

As early as shortly after the Dharamsala Earthquake of 1905, as reported by Dr A.S. Arya, a local guideline encouraging the use of masonry and timber infill frame construction like *dhajji dewari* was developed and promulgated. Now, like *taq*, *dhajji dewari* has been included in the Indian Standard Building Code *IS 4326, Earthquake Resistant Design and Construction of Buildings – Code of Practice.* It is identified as "brick nogged timber frame construction". In addition, just as with *taq* after the 2005 earthquake, the Ministry of Home Affairs, Government of India included *dhajji dewari* construction as one of the recommended types for new construction in their *Guidelines for Earthquake Resistant Reconstruction and New Construction of Masonry Buildings in Jammu & Kashmir State* (Arya and Agarwal, 2005, p.34).

Across the Line of Control in Pakistan Administered Kashmir, *dhajji dewari* (there referred to only as "*dhajji*") was formally recognized by the Government of Pakistan in November 2006, 13 months after the earthquake, as a "compliant building technique" that qualifies for post-earthquake financial assistance. During this period, many rural families had already used it to reconstruct their collapsed rubble stone houses, as is described in Chapter 5.

The quarterly report of the Earthquake Reconstruction and Rehabilitation Authority (ERRA), the *ERRA-UN Early Recovery Plan*, made what must be one of the most detailed and clearly stated case yet issued by a governmental body for the efficacy of *dhajji dewari* in rural construction (ERRA, 2006). The report states that:

> *Using many small panels and small elements, the building performs well in earthquake, distributing earthquake energy evenly, and damping the energy through friction. This relies on using infill that is not too strong or rigid...this differs from conventional frames, where the force is taken at the connections and corners.* Dhajji *construction should be understood as a system, and used as complete structures....Making* dhajji *compliant marks the first step on building on indigenous earthquake knowledge. From a technical perspective, for many affected communities* dhajji *construction, unlike the use of cement mortar, is easier to implement safely because it is not a new technology,... [and thus] is a move to safer construction while still allowing people to use familiar and recycled resources.*

The adoption of this language into the Pakistan government's report is significant because it not only shows that the traditional construction system was approved for the practical reasons that presented themselves after the earthquake in the rural areas, but that it was understood and embraced because of noteworthy technical features that give it good performance in earthquakes. In language that one would expect would be used to describe some modern high-tech seismic hazard mitigation device, *dhajji* construction is promoted with sophisticated points such as "damping the energy through friction...relies on using infill that is not too strong or rigid" and that "*dhajji* construction should be understood as a system, and used as complete structures". UN-HABITAT reported that by January 2009, over 100,000 new *dhajji* houses had been completed in the damage district (see Section 5.5 for more information).

3.5d & e *Full-scale model made by the Swiss Agency for Development and Cooperation (SDC) at the Housing Reconstruction Centre Balakot to show correct construction of the* dhajji *method for the training of carpenters and masons, spring 2007. Photo courtesy of SDC.*

WHY *DHAJJI DEWARI* CONSTRUCTION PERFORMS WELL IN EARTHQUAKES

1. Ability to sustain inelastic deformations without collapse.

2. Low initial stiffness of both systems reduces earthquake loading that otherwise would be destructive to masonry and lightweight timber joinery.

3. Large amounts of energy dissipation from friction between the masonry and timber and within the weak mortar joints of the masonry itself.

4. Timbers serve to stop the propagation of destructive cracking of the masonry.

3.6 A Comparison of Kashmiri Traditional Construction with the Provisions of the Indian Building Codes

While the early recognition of the value of these traditional systems and their inclusion in the codes is significant, it is important to note some of the differences and missing elements in the codes when compared with the best examples of Kashmiri traditional construction. This section is not intended to be comprehensive, but to draw attention to details that are distinct attributes of traditional construction and of critical importance to good performance in earthquakes.

It is important to remember that codes are intended to be minimum standards, so that the absence of a particular detail from any code does not mean that the code is deficient for its purpose, or that the detail itself should not be considered as a way of improving building performance beyond that of the minimum.

3.6.1 Floor-Level Ring Beam in *TAQ* and *DHAJJI* Construction:

IS 13827, IS 13827, Earthen Buildings and *IS 13828, Low-Strength Masonry* deal with bearing wall masonry, and the timber section of *IS 4326* which includes "brick nogged timber frame construction". Interestingly, all of these codes lack detailing of how the joists would be connected to the walls in multi-storey construction. *IS 13828* allows for construction of up to three complete storeys with a flat roof, or two storeys plus an attic in a pitched roof, but the illustrations and details showing the placement of the horizontal reinforcing bands show only one storey plus roof. For bearing wall masonry, the seismic bands in all of these codes are only shown as they would be for locations where the floor joists do not supplant the short connector pieces. In both the code sections on brick nogged construction and on timber frame construction, the method of connecting the floor joists securely to the walls is not illustrated or discussed.

As a result, the important feature of *taq* constructions where the floor joists extend through the wall between the timber runner beams embedded into the wall is missing from these codes. It is also missing from the 2005 *Guidelines for Earthquake Resistant Reconstruction and New Construction of Masonry Buildings in Jammu & Kashmir State*, which is based on those codes and allows two-storey plus attic gable construction in its provisions. This floor-to-wall connection detail is one of the most effective earthquake-resistant features of the *taq* and *dhajji dewari* systems. It

3.6.1 a, b & c Photographs showing construction details of taq *buildings. The one on the far left shows the front wall perpendicular to the joists in a* taq *building, showing how the joists are sandwiched between the timber bands. The second shows how the secure connection of the joists to the front wall bridged the earthquake-collapsed rubble-stone side wall. The third show a* taq *building partly demolished, exposing the timber lacing in the wall parallel to the joists.*

is essential for holding the building together and thus enabling the upper-floor and attic floor to work as diaphragms and resist the outward collapse of the exterior walls.

3.6.2 Corner Vertical Rebar in *Taq* Construction:

In *IS 13828*, low-strength masonry is defined as masonry laid in lime-sand or mud mortar. The code specifies that for this type of masonry a steel reinforcing rod should be placed vertically in the corner of the masonry walls, penetrating through the masonry and seismic bands from the foundation to the roof. It states: "Bars in different storeys may be welded or suitably lapped".

Of course, the traditional *taq* buildings which meet the definition of "low-strength masonry" were devoid of steel reinforcing or vertical reinforcing of any kind. The question that arises, therefore, is whether this added steel is a critical improvement over the traditional method of reinforcing only with horizontal timbers. While the vertical steel may seem to be only a small addition in the overall total scheme, it represents a large conceptual difference in the way that the building is structured not only to resist earthquake vibrations, but also to settle and move over time, as it must be allowed to do.

In a newly constructed one- or two-storey house, if well constructed with cement or cement-lime mortar, and with other steel reinforcement in the masonry, vertical rebars in the corners may very well raise the threshold at which the shaking would trigger damage; but the question of whether the construction will in fact be done well is a very important concern, and the code approves the use of clay mud mortar for both brick and stone masonry. In the fine print of *IS 13828*, the weak mortar and the existence of voids in rubble stone masonry is recognized by the recommendation that these vertical bars be embedded in a sheath of concrete by placing a "casing pipe" around them and lifting the casing as concrete is poured into it so as to confine the concrete to the area around the bar.

When this corner vertical reinforcement is understood for what it is – as a series of thin reinforced concrete columns arranged deep in the fabric of the load-bearing masonry walls of the one-to-three-storey house – one can see that the building that is being constructed is no longer truly a bearing wall building. This is because these tall thin columns with their steel cores cannot move and compress over time in a way compatible with the stone or brick laid in mud mortar that surrounds them. Instead, the steel bar has a different coefficient of expansion than the surrounding masonry, and, together with its concrete jacket, it cannot compress as the masonry and mud mortar around it shifts and compacts over time. Thus, it may gradually take a larger percentage of the overburden loads as the masonry and the mortar compress. This can gradually disrupt the masonry in the corner of the building.

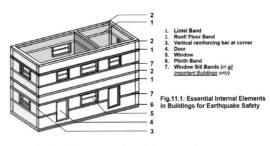

1. Lintel Band
2. Roof/Floor Band
3. Vertical reinforcing bar at corner
4. Door
5. Window
6. Plinth Band
7. Window Sill Bands (*in all Important Buildings* only)

Fig.11.1: Essential Internal Elements in Buildings for Earthquake Safety

3.6.1d This illustration from Ministry of Home Affairs Guidelines for Earthquake Resistant ...Masonry Buildings in Jammu & Kashmir State, by A.S. Arya. shows "seismic bands" as recommended for a two-storey building conforming to the Indian code. The seismic bands are important, but while much attention is given to vertical reinforcement of the corners (see Section 3.6.2), nothing is said about tying the floors to the walls through the floor-level bands. The floors are of major structural importance in earthquakes, and are critical to keeping a building from coming apart.

3.6.1e This is a section view of the house, the oblique view of which is shown in Figure 1.2d. This view provides the unique opportunity to see the entire dhajji *structural system at its most basic pre-industrial level. Few of the timbers have been sawn or shaped, and many are thus curved to the shape of the trees from which they came. The floor joists, which run from left to right, can be seen where they penetrate the exterior façade on the right between the top plate of the frame from the floor below and the bottom plate of the frame of the floor above.*

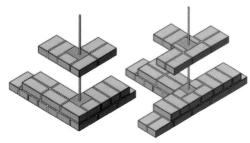

Fig.11.6: Recommended joint details with the vertical reinforcement at corner of brick masonry walls

3.6.2a Illustration from Ministry of Home Affairs Guidelines for Earthquake Resistant ...Masonry Buildings in Jammu & Kashmir State, *by A. S. Arya. This kind of detail is recommended in the Indian Standard Building Codes and other documents, but its long-term efficacy is questionable (see Section 3.6.2).*

3.6.2b This example of rusting steel demonstrates the expansion that occurs when the iron converts to an iron oxide (rust). The force of this on surrounding materials is sufficient to break any masonry or concrete cover.

In the event of an earthquake, the interface between this long thin column of the single steel bar in the concrete jacket can then be a zone of weakness in the critically important corners of the building because the column interferes with a proper strong bond pattern for the corner masonry between the timber bands.

In addition, cement is also incompatible with the mud mortars or weak lime mortars that are allowed in this same code. In the presence of moisture, the soluble salts in cement migrate out and crystallize, causing efflorescence that will gradually destroy the masonry and break down the surrounding mortar. Thus there is a significant danger of a rapid deterioration of the masonry walls. Moisture easily penetrates low-strength masonry. Normally this is not destructive, but when steel is introduced, the steel can rapidly become corroded in areas where the concrete sheath is cracked or inadequately filled when poured. This is inevitable in spite of the concrete sheath because the quality and consolidation of the concrete are likely to be compromised by the need to pack it into the small pocket, or even, in the case of rubble stone construction, pour it into the small casing tube.

The potential problems with this kind of detail are magnified by the code language that states that successive bars are to be "welded or lapped". Welding of steel reinforcing is dangerous, yet illustration no. 11 in the code shows a bar in a tube that is too small to accommodate the lapping of two bars with sufficient space for protective concrete cover. Welding changes the molecular nature of the steel, potentially causing brittle fracture of the rebar. It also can cause rapid corrosion because of changes to the electrical potential in the steel. Enlarging the tube to accommodate the lapping of each successive rebar will displace even more masonry, while failure to fully accommodate the lap with adequate concrete cover over it, and the off-centre bar above and below the lap will lead to rapid corrosion because of inadequate protection by the concrete.

It is important also to recognize that placement of both the rod and concrete jacket are skilled trades that are more advanced and less widely known than those of those of timber and mud-mortar masonry construction. Welding is an even more specialized skill. In most environments where the population continues to build using the traditional technology of low-strength earthen, stone, or brick masonry, it is unlikely that their knowledge encompasses proper safe construction in reinforced concrete, especially for such small components in what otherwise is a timber and masonry construction project. As a result, the inclusion of the vertical rebar in the code brings with it the substantial risk that, if executed in the field, the rural builders will only succeed in makinag their walls weaker, rather than stronger – even while adding substantially to the cost of the project.

Suffice it to say that if steel reinforcing had been available and used in the walls of the 19th- and early 20th-century *taq* houses, or for that matter the dome of the Pantheon in Rome, completed in AD 125 (Figure 5.2c), the bars would long ago have rusted, causing the walls to burst. Steel expands to many times its original dimension when converted from steel to iron oxide (rust), a process which exerts tremendous force onto confining materials (Figure 3.6.2c). Few of these buildings would be restorable today had this been done.

Ultimately, traditional construction typologies generally work best if one understands and thus maintains the integrity of each particular system rather than mixing it with potentially incompatible modern elements. This is particularly true for those traditional systems which have demonstrated structural attributes that have significantly reduced risk from different natural hazards such as earthquakes. The mixing of modern technologies, such as steel reinforcement inside a masonry wall, with traditional technology, such as the horizontal timber lacing in a low-strength masonry bearing wall, can radically change the behaviour of that wall in a way which can destroy the positive attributes of the older technology without effectively engaging the new. This is exactly what happened in Bam, Iran, where many of the houses collapsed during the 2003 earthquake, killing over 30,000 people. Almost all of these houses which collapsed were relatively new. Most had been constructed with unfired clay brick walls but had steel beams supporting heavy jack arch roofs of fired brick. These roofs were extremely rigid and strong, but lacked mechanical attachments to the unfired brick masonry walls, so they fell off the walls and crushed the occupants.

3.6.3 Height of timber-laced masonry buildings, including *TAQ* and *DHAJJI DEWARI*:

The timber construction section of Indian Standard code *IS 4326* begins with the statement that "timber has a higher strength per unit weight and is, therefore, very suitable for earthquake resistant construction". This positive comment is followed by a surprising restriction: "timber construction shall generally be restricted to two storeys with or without the attic floor". This is despite the fact that this type of combined construction has been common in many parts of the world for centuries, and has withstood earthquakes throughout the seismically active East European, Anatolian, and Central Asian regions. If timber is so suitable for earthquake resistance, is it not reasonable to ask: Why should timber and masonry buildings be limited to either one or two storeys? Two storeys is far less than what is allowed in other countries, including the United States, where timber construction even up to five storeys is

3.6.3a *A once splendid five-storey house of* taq *construction along side the Rainawari canal, Srinagar. These large houses in the Rainawari area are constructed on the very soft soils of what was once the middle of Dal Lake. The timber lacing is essential to their stability, and their age and durability demonstrates that multi-storey buildings of masonry and timber can be safely built.*

3.6.3b *A five-storey (including attic level)* taq *building and four-storey* dhajji dewari *building in the commercial centre of Srinagar. These and many others in Srinagar may be reminiscent of the kinds of buildings which astonished Kalhana in 1148, Tímúr in 1398 and Dughlat in 1540.*

3.6.3c The Gargoo Residence in Bulbul Lankar, Ali Kadal, Srinagar. This house is four storeys plus attic in taq *construction, with a stone base and brick upper storeys. Photo by Akshay Kaul.*

common, and where historically it was often higher (see Figures 3.6g and 3.6h). If fire is the concern, the brick infill in brick nogged timber construction tends to slow down the spread of fire compared with the 100% timber construction with pocket walls that is common in North America. If maintained, wooden structures can last for hundreds of years (see Figures 5.5a & 5.6e).

On the issue of building height, it is worth noting the early observations quoted in the previous chapter from 1398 by Tímúr who observed that the buildings in Srinagar were "all of wood" and "they are four or five storeys high" (Elliot, 1867) and the other by Mirza Haider Dughlat (1540) who remarked on the "lofty…five storeys high buildings of fresh-cut pine" (Bamzai, 1994). If 16th-century timber buildings could exceed five storeys before there was modern firefighting apparatus, then why only two storeys now?

In any case, as remote as the older buildings in Srinagar and Baramulla seem from those that are illustrated in the Indian Standard Building Codes, it is still important to understand that these older buildings are part of a living tradition. The reason why the elements of their construction are in these codes is that certain scientists and engineers, with a knowledge of more than only steel and concrete, recognized aspects of these premodern systems as earthquake-resistant long before the 2005 earthquake. Perhaps it is now time to consider embracing the use of traditional materials and systems more widely and wholeheartedly, so that the full benefits of sustainable and environmentally less destructive construction practices can be obtained.

Building Tall with Timber in Other Parts of the World

Before the advent of reinforced concrete and steel, timber and masonry were the materials of choice for multi-storey structures. Now in earthquake areas, building regulations have often restricted the height of such buildings – sometimes even to only a single storey. As history can demonstrate, this is over-conservative, even in earthquake areas.

3.6.3d & e Five- to seven-storey buildings in 17th- to19th-century Madrid, Spain, were constructed entirely, except for the street facades, with timber frames with masonry infill. In 3.6.3e, demolition revealed this little-known construction feature that one time was common throughout the city.

3.6.3f The massive substructure of Kyoto's Kiyomizu Temple was constructed in the 17th century, and stands equivalent to a ten-storey building, even when disregarding the additional height of the temple on top of it.

3.6.3g Many 19th-century hotels in North America, like the ten-storey 279-room Claremont hotel in Oakland, California shown here, were constructed entirely of wood.

3.6.3h This new five-storey central city residential complex in Oakland, California shown under construction in 2008 is almost entirely of timber stud frame construction. It covers four city blocks.

3.6.3i With an area of 340 x 115 metres, and equal in height to a 15-storey building, this and seven other surviving World War II dirigible air docks are the world's largest wooden structures. Each is 12 times larger than Japan's 18th-century Todaiji Temple, which is often claimed to be the world's largest wooden building.

3.6.3d Historic centre, Madrid

3.6.3e Rear of buildings in 3.6j

3.6.3f Kiyomizu Temple substructure

3.6.3g Claremont Hotel, Oakland, Ca.

3.6.3h New apartments, Oakland, Ca.

3.6.3i Hanger 2, Moffet Field, California

4a This taq house in Srinagar, cut in half to allow for a bridge widening, shows well the arrangement of the timber lacing in the walls.

The Repair and Strengthening of
Taq and *Dhajji Dewari* Buildings

Traditional structures passed the test. In Kashmir…traditional timber-brick masonry…has excellent earthquake-resistant features…Its performance in this earthquake has once again been shown to be superior, with no or very little damage…Traditional skills could be called upon in the effort to create sustainable communities in the face of the risk of further earth tremors.

Building Talk, Chartered Institute of Building (CIOB)
Hertfordshire, UK, 2006

In the prior chapters, the merits of traditional construction have been described, but of course this message must be tempered by the realization that not all buildings are of the same quality or condition, regardless of the materials or construction system. While traditionally constructed buildings may be more forgiving than reinforced concrete as regards worker training and construction quality control, they are not immune to the effects of poor construction, deterioration, and/or lack of maintenance.

Being a biological material, wood is prone to both fungal decay and insect attack. In Kashmiri buildings, structural timbers are often exposed to the weather on the exterior of the buildings, which can increase the vulnerability to decay if not properly maintained. If sapwood rather than heartwood has been used, vulnerability to decay is increased. This is also true if fast growing short-lived trees have been used rather than slower growing longer-lived trees, for reasons explained below. Since timber is one of the key components of their earthquake resistance, rotted timbers can radically reduce the earthquake resistance of a *taq* or *dhajji dewari* building.

While weaknesses from environmental deterioration may be hidden, any damage that has occurred after an earthquake will usually be visible, yet the structural effects of such damage may not be easily understood. Because the earthquake resistance of traditional buildings is a direct product of their inelastic behaviour and energy dissipation, there may be extensive amounts of fallen plaster and other disruption, yet this does not necessarily mean the structure has lost a significant amount of its earthquake resistant capacity. By contrast, small cracks in reinforced concrete structures, depending on where they are, may be indicative of much greater vulnerability to collapse than even larger cracks in traditional buildings.

For example, when a large downtown reinforced concrete office building, the five-storey Rubén Darío Building, collapsed in the 1986 San Salvador earthquake, killing over 200 people, it was revealed it had been damaged in a 1965 earthquake sufficiently to have been condemned at that time by building officials, but it had continued to be used for another two decades without having been structurally repaired (Anderson, 1986). Apparently, the serious structural damage was inconspicuous enough to make it possible to simply ignore or cover it over and go on using the building. It is therefore difficult, but important, to be able to understand the difference between damage which puts a building at risk, and that which does not. As with most things, not all earthquake damage is equal.

This chapter will describe how to approach the maintenance and repair of buildings with both types of traditional construction. The chapter is divided into four parts. The first three parts deal with the long-term maintenance and repair of *taq* and *dhajji dewari* structures, and the fourth part focuses on post-earthquake damage assessments, and the problem of obtaining proper evaluation and treatment of buildings with traditional structural *taq* and *dhajji dewari* structural systems. This division of the chapter can be characterized as (1) what to do now, before the next earthquake, and (2) what to do after the next earthquake, with reference to experiences from the 2005 earthquake and other earthquakes. The sections that follow are:

4.1 Assessment, Repair and Strengthening of Structural Materials

4.2 Structural Integrity Analysis of *Taq* and *Dhajji dewari* Systems

4.3 Evaluation and Repair of Structural Conditions Resulting from Alterations and Additions

4.4 Post-Earthquake Building Inspection and Evaluation

The first section covers the basic structural materials of wood, masonry and concrete. The second section is focused on the *taq* and *dhajji* structural systems as a whole, in order to provide advice on how to deal with problems associated with the maintenance and repair of traditional Kashmiri buildings. It also covers how to deal with anomalies that may compromise the integrity of the structure in individual buildings.

The third section may be the least understood, but it addresses a very common yet difficult problem: the alteration of traditional flexible constructions with rigid components. These may be (1) upper-floors of reinforced concrete or masonry construction, (2) adjacent wings of reinforced concrete, unreinforced masonry, or reinforced masonry construction, or (3) internal modifications such as the addition of partition or exterior walls of masonry in cement mortar, or concrete topping slabs poured over the existing floors. Concrete block masonry construction and reinforced concrete floor slabs are extremely rigid, and thus are incompatible with the flexibility of traditional constructions of any type, but in combination with *taq* or *dhajji* construction, non-engineered additions or renovations utilizing stiff and heavy elements of concrete or unreinforced masonry will conflict with the flexibility of the traditional structural systems, causing more damage than would otherwise be likely to occur.

These first three sections also contain some recommendations for strengthening of traditional timber and masonry structures. The case made here for the general good behaviour of *taq* and *dhajji dewari* buildings in earthquakes does not mean that the resistance of any particular traditional structure cannot be improved or that strengthening procedures should not be considered. There may be important deficiencies in individual buildings that need to be addressed. The recommendations are intended to provide a preliminary guide to some of the interventions that can be beneficial.

4b The interior of the second storey of a Turkish hımış *construction house after the 2000 Orta earthquake, showing the extensive disruption to the plaster, mainly along the interface between the masonry infill and the timber frame. The shaking from this earthquake was less than in the great 1999 earthquakes.*

What turned out to be significant was that, while reinforced concrete buildings in Orta suffered what appeared to be a small amount of damage, these same types of buildings collapsed by the hundreds in Gölcük and Adapazari in 1999.

However, the damage to hımış *structures in each of the two earthquakes was very often similar. The reinforced concrete buildings proved to lack the reserve capacity that the traditional structures demonstrated (see Langenbach and Gülkan, 2004 for more information).*

The fourth section deals specifically with the problems that have been encountered in past earthquakes where damage to traditional buildings has been misunderstood, leading to unnecessary displacement of people and loss of cultural heritage. Recent earthquakes where these problems were evident include the 2005 Kashmir earthquake, the 2001 Gujarat earthquake, and the 1999 and 2000 earthquakes in Turkey. It is important to understand that in most cases, *taq* and *dhajji dewari* buildings can be restored and do not need to be demolished even when damaged by an earthquake. However, before the next earthquake occurs, an effort must be made to check the load path and structural integrity of individual buildings as described in Section 4.2, as well as the material condition (Section 4.1) and the impact of later alterations (Section 4.3), and to correct any deficiencies. In all cases, a qualified professional competent in traditional construction methods should be consulted before undertaking any work.

These sections are intended to provide a basic understanding of the elements that contribute to safe conditions in Kashmir's vernacular buildings. Owners are encouraged to seek further assessment from professional architects and engineers with specific knowledge, understanding of, and sympathy for, traditional construction. "Sympathy for" is included in the last sentence because most professionals practising now have been largely trained to use reinforced concrete, and they may carry a prejudice against relying on traditional construction in earthquake areas. When this is so, they will not be able to provide good advice.

Architects and engineers undertaking more detailed analysis are encouraged to consult other international resources that may be of assistance in undertaking their analysis, even though they may not be specific to the Kashmir construction or environmental conditions. Some of these can be found on the web on the "Library" page at www.traditional-is-modern.net.

For structural engineers and architects, two documents available in their entirety on the "Library" page at www.traditional-is-modern.net are useful for analysing and strengthening a wide variety of existing building types. These are (1) *FEMA 310: A Handbook for the Seismic Evaluation of Buildings, A Prestandard* and (2) *FEMA 356: A PreStandard and Commentary for the Seismic Rehabilitation of Buildings*, produced by the Federal Emergency Management Agency, the US government agency charged with setting code standards for disaster mitigation. These documents are now widely used internationally. A Basic Checklist for structural engineers from *FEMA 310*, containing those items most relevant to the traditional buildings in Kashmir, can be found at the end of Section 4.3 below. Although the conditions in Kashmiri buildings are quite different, this checklist is relevant enough to merit its inclusion in this volume.

4c A view of Patan, Kathmandu Valley, Nepal, after the 1934 earthquake shows the collapse of masonry structures that lacked complete timber-laced systems. See Section 4.2, Building Integrity Analysis for Taq and Dhajji Dewari Systems. Source: Proksch, 1995.

4d This picture shows a collapsed masonry wall following the 1906 San Francisco (USA) earthquake, which displays the vulnerability of masonry walls that are not properly tied to floors. See Section 4.2, Building Integrity Analysis for Taq and Dhajji Dewari Systems. Source: San Francisco Museum.

4e Five-storey taq *building in Srinagar showing the timber lacing at the floor levels, with the floor joists sandwiched between the timber runners. The house has a substantial tilt from differential settlement, but the timbers are critically important in keeping the house from splitting. See Section 4.2, Building Integrity Analysis for* Taq *and* Dhajji Dewari *Systems..*

4.1 Assessment, Repair and Strengthening of Structural Materials

4.1.1 Wood: Is there evidence of decay (rot or insect attack) to the structural timbers?

CONDITION ASSESSMENT: The first priority in any condition assessment is to inspect the structural timbers in a building. This includes not only the interior beams and columns that can be readily seen, but also the runner beams in *taq* walls and the members that make up the *dhajji dewari* frames. Most critical is to inspect the condition of beams where they are in contact with masonry, or which penetrate through to the exterior. It is also important to inspect the connections between timbers, both to verify what fasteners have been used, and to determine if they have deteriorated.

Any timbers deteriorated from insect or fungal decay can compromise the structural safety of the entire building. Timber loses much of its structural capacity if fungal decay or insect attack is advanced, yet sometimes this damage is not readily apparent. The cut ends of timbers are more vulnerable to decay because the moisture is more easily absorbed by the wood through the severed ends of the cellulose fibres.

> TIMBER LOSES MUCH OF ITS STRUCTURAL CAPACITY IF FUNGAL DECAY OR INSECT ATTACK IS ADVANCED, YET SOMETIMES THIS DAMAGE IS NOT READILY APPARENT.

Care must be exercised to inspect timbers that extend into masonry walls or penetrate to the exterior where fungal decay can gain a foothold on the exposed ends. It is particularly important to inspect evidence of decay in any wood where the end grain is in contact with masonry or concrete, as this would compromise the attachment of the floors to the walls. In all but the most extreme cases, this problem can be repaired by replacing affected timbers. If rotted beams or columns are reinforced with new timbers, rather than replaced, care must be exercised to prevent the fungus from infecting the new timbers by isolating the timbers from each other, and treating the fungus with chemicals (see below) or removing the infected parts. The source of the moisture needs to be addressed as well.

4.1.1a (left and middle) This hımış *construction house in Orta, Turkey was photographed the day after the 2000 Orta earthquake caused it to collapse. It had long been abandoned, and the timbers were rotten. All the occupied and maintained houses of similar construction in the same village remained intact. (Right) Rotted timbers at the base of the framing members of another* hımış *construction house near to Orta, Turkey. The abandoned house remained standing in the 2000 earthquake despite this damage, but direct wood-to-ground contact can easily lead to rapid deterioration of timber structural members, resulting in dangerous conditions.*

REPAIR GUIDELINES: The dark wood at the centre of the trunk, the heartwood, is older and more durable than the younger wood surrounding it. As a tree grows, a thin layer of cells called the cambium generates new wood, called sapwood, just under the bark. Sapwood is softer and tends to be lighter in colour than heartwood. As the sapwood ages, natural substances called extractives invade it and gradually convert it to heartwood. It is these extractives that protect the wood used in construction from insect and fungal attack. Certain tree species such as Himalayan cedar called *deodar* (from Sanskrit, meaning "timber of the gods") have more resistant heartwood than others, and local knowledge and experience should be used to make this choice. In general, the species that live the longest have the greatest resistance to rot and insect attack when harvested for building materials.

- Learn to identify deteriorated wood and the agents of wood decay. Wood decay is caused by fungus and/or insects, both of which are caused and aggravated by the presence of excessive moisture.

- Repair sections of wood, if compromised by rot of insect damage. Chemicals can be used to resist or suppress fungal and insect attack, but care should be exercised, as they can be toxic to people and to the environment. Boron compounds (Borax, Boric Acid, etc.) and Propylene Glycol (used for common antifreeze – but avoid Ethylene Glycol) are proven to have low toxicity, are inexpensive by comparison to more toxic solutions, and have been proven effective; however, it is important to undertake research in Kashmir on availability of particular biocides and on their comparative effectiveness on the different species of construction timber found in the houses.

- For structural work on exterior and above foundations, use only heartwood of a species of timber with proven insect and rot resistance, or that has been chemically treated to enhance resistance.

- Never use sapwood of any species on the exterior or where it is in contact with masonry or concrete, such as on foundations or inserted into masonry walls, or where it is subject to frequent wetting or high humidity.

- Repair sources of moisture penetration promptly, and increase ventilation in areas subject to condensation.

- Protect the ends of timbers where they will be subject to frequent wetting. Wood absorbs moisture more easily through its end grain than from the side. Avoid direct unventilated contact between the ends of timbers and masonry or concrete.

- Avoid the spreading of fungal decay (wood rot) and insect attack from infected to uninfected wood. Take care to remove fungus-infected wood from contact with new or sound wood. Dispose of insect-infested wood safely off-site. For example, it is unwise to stack rotted wood for stove use in contact with timber houses.

- Do not replace wooden structural members in *taq* or *dhajji dewari* structures with any other material of different properties, such as concrete, which is much stiffer than the wood members being replaced, and which will attract condensation moisture into remaining parts of the wooden structure.

- Augment or replace rusting nails or other deteriorated timber connections. Avoid the use of unprotected bare steel nails for permanent structural work, particularly on the exterior. Instead, use galvanized or stainless steel nails for exterior construction.

- Avoid the use of waterproof sealants or cement plaster on exterior wooden and masonry elements, as these can trap moisture behind and cause more harm from the lack of ventilation.

STRENGTHENING RECOMMENDATIONS: Strengthening of timber connections and connections of the timbers to the masonry walls, particularly at the corners of a building, and securing the walls to the floors can significantly improve earthquake resistance. This is best done in conjunction with the replacement of any decayed timbers. Re-nailing of *dhajji* timber frames with new stainless steel nails (best, but expensive) or galvanized nails (good) is wise because nails which cannot be seen may rust over time, resulting in considerable loss of strength.

NOTE: Avoid using reinforced concrete for retrofitting new seismic bands above the *das* into existing historical *taq* buildings, as this can do more harm than good. Generally in buildings of traditional construction, concrete is suitable only for work at the foundation level.

IDENTIFICATION OF SAPWOOD AND HEARTWOOD

4.1.1b Sapwood is softer and tends to be lighter in colour than heartwood. It is only the heartwood (the red colour in this example) that is resistant to fungal decay and insect attack.

(D)   (E)

Evidence of insect attack.

(F)   (G)

Simple test for structural integrity of timber.

4.1.1c

(A)

(B)

(C)

Evidence of structural decay.

IDENTIFICATION OF WOOD DECAY

Periodic inspection for decay is critical to the safety and habitability of homes. Evidence of decay can sometimes be obvious, and at other times less so. In (A) moss growing on the surface is a strong indication that the wood itself has been attacked by wood-consuming organisms. In (B) and (C) the evidence is not so prominent, yet wood fungus has structurally damaged the member and then died back as the wood dried out. The evidence of decay that is sufficient to cause a loss of strength is the cracking of the side and end grain surfaces into a rectangular pattern. This is the signature of fungal decay after it has dried out. The wood is partially consumed, and then the seriously weakened remaining tissue dries and cracks.

Wood-destroying fungi tend to survive through dry spells, so that when moisture returns, the fungus will reemerge. Thus it is important to treat or remove wood that shows evidence of fungal decay.

The evidence of insect attack can be seen in (D) and (E). The telltale marks may only be the small round holes visible in (D), but if left unchecked, the interior of the timber will eventually be consumed, as seen in (E), sometimes with little evidence of the amount of hollowing-out visible on the surface.

A simple test for damage from fungal decay is to use an awl or other sharp tool to evaluate the fibre strength of the timber by breaking the surface. Short splinters that break apart easily from this test as seen in (F) are evidence that the wood has lost its structural capacity, whereas long splinters from this test as seen in (G) are evidence the wood is sound. Extensive damage from insects can sometimes be located initially by rapping on the timbers with the handle of a tool. If a hollow sound is produced, this is an indication that the timber is no longer sound.

4.1.2 MASONRY DAMAGE AND MORTAR DETERIORATION: ARE THERE CRACKS IN THE MASONRY OR SECTIONS OF THE MASONRY WALLS THAT HAVE COLLAPSED OR FALLEN OUT?

FOR THE REPAIR OF MASONRY, IT IS BEST TO USE THE SAME KIND OF MORTAR THAT WAS USED FOR THE ORIGINAL CONSTRUCTION.

CONDITION ASSESSMENT: It is very likely that the masonry walls of many, if not most, of Kashmir's historic vernacular buildings have not been maintained or repaired in many years. Considering this length of time, it is remarkable that more deterioration has not occurred than is clearly evident now. The repair of masonry is a specialized skill that cannot be described in detail within the limits of this publication, so the recommendations here are intended to highlight the importance of retaining local experts who are familiar with the historic masonry, not just with modern work which is entirely different, and also to warn of the dangers of undertaking such work by reaching for the conventionally used bag of cement mortar.

REPAIR GUIDELINES: For the repair of masonry, it is best to use the same kind of mortar that was used for the original construction. If this is not done, the repairs can cause rapid deterioration of pre-existing work around the repairs. Mud mortar has been used historically in Kashmir most often, but stronger lime mortar exists there as well. Lime mortar is widely recognized as an essential ingredient in good masonry mortar in restoration work in many countries.

Unfortunately, lime for mortar is not now generally available or used in Kashmir, and so it will be important to revive its use and make it available. Good quality masonry work is dependent on it and it is important that it be reintroduced in Kashmir for restoration work as an alternative to the now ubiquitous use of Portland cement. Portland cement is not a good substitute for hydrated lime (non-hydraulic lime). Its use on traditional buildings is destructive because it is too rigid and brittle and because it introduces soluble salts into historic masonry that cause rapid deterioration. As a result, it will cause damage in flexible *dhajji dewari* walls and deterioration of the masonry in *taq* walls.

Generally, walls in *dhajji dewari* buildings are thin enough to repair by repointing, but *taq* walls are built from many courses of brick, and their repair will sometimes require consolidation with a hydraulic mortar. This is a very specialized process, involving proper research on the best available ingredients for the purpose, but the most important recommendation that needs to be made here is that ordinary Portland cement mortar be avoided, as it can cause irreversible damage to traditional mud and lime mortared masonry. There are formulae for lime-based grouts that can take on a hydraulic set (harden on their own without CO_2 from the atmosphere), but which avoid the disadvantages of ordinary Portland cement.

- Use mortar with the same or compatible constituents as the original mortar.

4.1.2a Detail of the corner of an abandoned house in Srinagar in ruinous condition vividly shows the historic masonry in taq *construction. The hard-fired "Maharaji" bricks are only the external skin, and the interior is random rubble of clay and low-fired bricks. Repairing this kind of damage is complicated by the need to integrate the new work with the old, rather than introducing a stiff hard block of new masonry that cannot work together with the historic fabric.*

- Do not use any mortars that are stronger than the masonry units.

- Lime mortar can be considered an alternative to plain mud mortar for fired brick and stone masonry.

- Do not use Portland cement in masonry mortar for repointing, repair or reconstruction of any traditional masonry wall. Avoid the use of Portland cement, except in below-ground foundations.

- Never add load-bearing unreinforced masonry walls on top of *dhajji* walls.

STRENGTHENING RECOMMENDATIONS: Even though both *taq* and *dhajji dewari* construction have floor joists sandwiched between the timber bands, an important structural asset described in Section 3.6, most buildings can be improved with the addition of wall anchors connecting the walls parallel to the joists to the floors to ensure against the outward collapse of the walls (see Figure 4d). This is more critical with the *taq* and *bhatar* buildings because of their heavy walls. Retrofitting should be specified by an engineer or architect with specific knowledge and experience with unreinforced masonry buildings, and with timber-laced masonry buildings.

4.1.2b *A masonry house under construction, probably unreinforced, photographed in October 2005 in Srinagar, one year after the earthquake. It appears that the lower-storey walls are laid in mud mortar, while cement was used for the mortar above. Already, the mud-laid lower storey walls appear to be coming apart.*

4.1.2c & d *This is a close-up view of a new house of* dhajji dewari *construction photographed in the 1980s. The house can be seen in Figure 5.5b. The* dhajji *walls were correctly being constructed with mud mortar, shown being mixed, but for some inexplicable reason, a back addition was being construction at the same time without the* dhajji *frame. For this, cement mortar was being used, which would be acceptable, except that in the event of an earthquake, the likely incompatibility in stiffness between the two building parts could cause damage.*

4.1.3a The exterior of this dhajji house in Pakistan Administered Kashmir was covered with cement plaster after the 2005 earthquake. This plaster layer cracked and delaminated in the 20 February 2009 tremor. Once cracked, it is difficult to repair. Photo by UN-HABITAT.

4.1.3b This ferro-cement plaster layer, installed on a building in Italy, blocked the normal flow of moisture from condensation. This in turn has caused the reinforcement in it to rust, as well as causing deterioration of the underlying stone masonry. It then delaminated during the Umbria-Marche earthquake of 1997. Photo by Sergio Lagomarsino.

4.1.3 PLASTER: IS THE EXTERIOR OF THE BUILDING CURRENTLY COVERED WITH PLASTER, OR IS NEW EXTERIOR PLASTER BEING CONSIDERED?

CONDITION ASSESSMENT: The unique aesthetic and historic quality of vernacular architecture in Kashmir is the texture of its exposed masonry and timber walls. Most traditional urban Kashmiri buildings have exposed masonry on the exterior. In Kashmir, the climate with a hot and damp summer and a cold winter demands that a wall be able to breathe both ways. Water can attack walls not just from rain, but also from condensation of interior water vapour onto cold surfaces. Traditional construction materials often perform better in this respect than many modern materials, particularly concrete.

REPAIR GUIDELINES: It is recommended that clay or lime plasters be used for plastering walls. Traditional plasters made of lime and gypsum "breathe", that is, they allow the natural transmission of water vapour. Do not use Portland cement plaster on either the exterior or interior of *dhajji*, *taq* or *bhatar*, or masonry buildings of any kind. Cement used in traditional buildings can be very destructive for a number of reasons: (1) it creates an impermeable layer that promotes the deterioration of both the wood and the masonry; (2) even though it is strong, it is brittle and easily cracked, as seen in Figure 4.1.3a; (3) once cracked, it can cause water to enter the wall; (4) it introduces soluble salts into the walls, destroying the masonry. Because the cement is impervious to water and water vapour, the trapped moisture will quickly deteriorate the fabric of the building.

STRENGTHENING RECOMMENDATIONS: A number of manuals offering post-earthquake guidance have recommended strengthening buildings by installing a ferro-cement layer over masonry. This has become a common practice in a number of countries.

In the case of traditional masonry buildings in Kashmir or any other areas of India and Pakistan, including *taq* and *dhajji dewari* buildings, this practice should be avoided as it is potentially extremely destructive of the masonry. Not only does it ruin the architectural character of the buildings, in an earthquake it initially will make the buildings more rigid, which attracts more forces. When the ferro-cement layer cracks and separates from the underlying *dhajji* or *taq* construction, it can cause unpredictable results such as torsion or localized fracture or collapse of the underlying masonry leading to a soft-storey condition.

Where additional strength is needed, it is recommended that additional *dhajji* walls be constructed on the interior, or, for *dhajji* buildings where the loads are determined to be too large for a given wall, a second *dhajji* wall can be installed against an existing wall. This kind of work is much more compatible with traditional masonry than reinforced concrete, including steel-reinforced cement plaster.

4.1.4 Cement and Concrete: Where are cement and concrete appropriate to use?

BACKGROUND: This book frequently cautions against the use of Portland cement concrete. However, because of its unique properties, it is the best material to use for some repairs and upgrades.

Cement was discovered by the Romans when they noticed that volcanic ash (pozzolan) when mixed with hydrated lime would set under water. Reinforced concrete came into use beginning in the 19th century. Concrete is a construction material composed of cement (commonly Portland cement) as well as other cementitious materials such as fly ash and slag cement, aggregate (generally a coarse aggregate such as gravel limestone or granite, plus a fine aggregate such as sand), water, and chemical admixtures. The word concrete comes from the Latin word *concretus*, which means "hardened" or "hard".

Cement is a binder. The word cement derives from the Roman term *opus caementicium* describing masonry which resembled concrete which was made from crushed rock with burnt lime as binder. The volcanic ash and pulverized brick additives which were added to the burnt lime to obtain a hydraulic binder were later referred to as *cementum, cimentum, cäment* and *cement*. The term "Portland cement" came from the resemblance of 19th century cement developed in England to the limestone from the Isle of Portland.

"Hydraulic" refers to the fact that it will harden from the chemical reaction between its components and water, a process called hydration. This does not require CO_2 from the atmosphere, as is required for lime mortar to set. The water reacts with the cement, which bonds the other components together, eventually creating a stone-like material. As of 2007, about seven billion cubic metres of concrete are made each year, more than one cubic metre for every person on earth.[12]

Reinforced concrete has steel reinforcement around which the wet concrete is placed. Once hardened, the combination has the compressive qualities of concrete combined with the tensile and ductile properties of the steel, making for a useful structural combination.

REPAIR AND STRENGTHENING RECOMMENDATIONS: For repair and maintenance of heritage buildings, particularly buildings of traditional timber and masonry construction, concrete can be a very useful material for use in foundations and other parts subject to ground water penetration or contact. For some buildings where the lowest level timbers *(das)* have rotted out, it is possible with careful and creative construction work to devise a substitute reinforced concrete bond beam to replace it.

It is important to note that the steel used for reinforcement of concrete must be fully and adequately covered by the concrete without rock pockets or air gaps, or it will quickly corrode, which then will break up the concrete. Thus, workers placing concrete must be

SINCE THE *DAS* IS IN CONTACT WITH THE FOUNDATIONS OF THE BUILDING IT CAN BE THE ONE EXCEPTION TO THE RECOMMENDATION AGAINST THE USE OF CONCRETE FOR REPAIR WORK.

4.1.4a A concrete mixing lorry (transit mix) in Sultandagi, Turkey. A transit mix is necessary to obtain consistent good quality mix for all but very small amounts of concrete.

adequately trained. Being certain of adequate consolidation of the concrete, and coverage of the steel rebars, in retrofit situations such as *das* repair is very difficult, so considerable skill and care is needed in the construction work.

It is not advisable to use reinforced concrete to replace timber members higher up in a *taq* building because of the differences in dynamic characteristics that are described below in Section 4.3. Since the *das* is in contact with the foundations of the building, and often ends up being partly covered with soil at the ground level, it can be the one exception to the recommendation against the use of concrete for repair work, but it should be evaluated on a case-by-case basis.

4.1.4b The steel rebars in this newly poured reinforced concrete column for a new mosque in the earthquake-damaged town of Sultandagi, Turkey remain exposed because of poor consolidation of the concrete at the time it was poured into the formwork. This condition, if not repaired before the building finishes are installed, will lead to corrosion, but repairs can never make up for the absence of strength and ductility caused by such poor consolidation.

With proper inspection for code compliance and quality control, such work should require demolition and replacement. Had a mechanical vibrator been used, this would not have happened. (See Section 5.3 for more information.)

4.2 Structural Integrity Analysis of *Taq* and *Dhajji Dewari* Systems

4.2.1 Foundation to Roof Structural Integrity: Does the *Taq* or *Dhajji Dewari* Construction Extend from Foundation to Roof Level Alone or in Combination with *Dhajji Dewari* above the *Taq*?

Condition Assessment: *Taq* construction often extends to the eaves of the lowest part of the roof, with *dhajji dewari* used to infill the gable ends of the roof. Sometimes *taq* is used only for the lower floors, with a full floor or more of *dhajji dewari* above. Since *dhajji* construction is lighter than *taq* and does not depend, as does *taq*, on a certain amount of overburden weight for its lateral resistance, the use of *dhajji dewari* above *taq* is good earthquake-resistant practice.

Although whole buildings of *dhajji dewari* construction are common, it is frequently used for the upper storeys over *taq* or unreinforced masonry construction below. The combination of *dhajji dewari* with *taq* is earthquake resistant because the lightweight infill frame construction helps to pre-stress the bearing wall masonry below, while holds the upper part of the building together. However, as was revealed in a number of instances in the 2005 earthquake, the absence of any timber lacing in the lower floors of the structure can lead to a collapse of those walls out from under the *dhajji dewari*.

Repair and Strengthening Recommendations: It is especially critical that the top-level diaphragm (the floor of the attic) be securely connected to the walls because there is little overburden weight at that level. Mechanical ties are necessary.

- Anchors between the floor diaphragms and the walls should be added if the secure attachment between the floor and wall timbers is lacking or compromised. This is the most important first step.

- In conjunction with wall ties, improvements to the diaphragm of the top (attic) floor can be made by adding nails to the floorboards, or to wood board ceilings. Although heavy, an existing floor of well consolidated earth and straw between the joists may work, but replacing this with poured concrete should always be avoided.

- Any masonry eaves, parapets, or other architectural elements that could break away from the structure and fall need to be secured.

- New floors of reinforced concrete should never be placed on top of *dhajji* walls and concrete topping slabs should never be added on top of wooden floor structures.

- The timber framework or timber beams in masonry should not be replaced with reinforced concrete (with the exception mentioned in 4.1.4 above.)

- Do not insert vertical reinforcing into the masonry walls in *taq* construction (see Section 3.6.2 for an explanation).

The use of *dhajji dewari* above *taq* is good earthquake-resistant practice.

4.2.1a Collapse of unreinforced masonry walls below left the dhajji *wall suspended above. The lower part of the walls lacked ties securing the floors to the walls and timber lacing in the walls. (See also Figure 4d for a similar example from the 1906 San Francisco, USA, earthquake.)*

4.2.1b A commercial building in Srinagar with the first two floors of taq, *and the top floor in* dhajji dewari.

4.2.1c This example of a small building in Srinagar photographed in 1981 shows how radically out-of-plumb timber-laced masonry buildings can be while remaining intact.

A NOTE ABOUT THE LEANING BUILDINGS OF SRINAGAR: SOIL SUBSIDENCE

Because both *taq* and *dhajji dewari* buildings are capable of remaining standing despite significant amounts of differential settlement, a leaning or deformed structure should not be a reason alone for condemnation. In fact, one of the important historical reasons for the local adoption of both systems is that, unlike unreinforced masonry, they are capable of significant deformation without being weakened or losing stability.

It is, of course, important to monitor building deformation to detect if there is on-going movement that could lead eventually to structural failure. In a case where the lean has shifted the building's vertical loads significantly off-centre, as is the case in the rather extreme example shown in Figure 4.2.1c., it is important to analyse the structural stability of a structure, and install buttresses or bracing as necessary.

For a visible deformation to happen without causing large cracks, it is of critical importance that the mortar used for laying or repointing the masonry be of low strength (mud or lime), rather than high strength (cement mortar). With the timber lacing and soft mortar, the stresses in the masonry are easily relieved as the masonry readjusts to conform to any deformed shape caused by differential settlement.

4.2.1d This taq *building in Srinagar shows how the timbers can serve to hold the masonry together even when the wall becomes seriously deformed by differential settlement.*

4.2.2 Floor to Floor Structural Integrity: *Taq*: Do the timber ladder-bands exist at (1) the floor (or plinth) level of each floor, (2) the window lintel level, and (3) the roof eaves level?

CONDITION ASSESSMENT: In *taq* construction, the common locations of the bands is at the floor levels and at the window lintel levels, but some houses, particularly with arch-top windows, have them only at the floor levels. This is not as resilient, but is still considerably better than unreinforced masonry without timber lacing. The Indian Standard codes recommend "seismic bands" (which include timber bands in certain types of masonry construction) at the window and door lintel level, and at the floor level for each floor. *Bhatar* construction sometimes is constructed with dry-laid masonry consisting of irregular broken stones. In these cases, the best results will be if the timber bands are relatively close together, and the stone used is freshly quarried and flat, rather than rounded river rock, and the wall has horizontal bedding planes and bond stones through the wall.

4.2.2 This cross-section of a partially demolished wall was taken in Ahmedabad in 2006. As can be seen here, the timber bands were more elaborate than usually found in Kashmir, with double rows of runner beams even at the mid-wall height level, as well as at the floor levels. See also the view from above in Figure 1.1h, and the detail of the cross-timber shown in Figure 2.3.3a.

The timber-laced brick masonry construction found in the historic walled city of Ahmedabad shares the same technology as Kashmiri taq *construction. Many of these buildings, as can be seen in Figures 3.2c & 3.2d, also utilized the timber lacing as part of the ornament.*

THE MAXIMUM SPACING
OF TIMBER FRAME
ELEMENTS FOR INFILL
MASONRY WALLS SHOULD
NOT EXCEED ONE METRE
BETWEEN STUDS AND
ONE METRE BETWEEN
HORIZONTAL MEMBERS.

4.2.3a This mixed taq *and* dhajji *building in Baramulla suffered earthquake collapse of the stone wall because the timber band at the lintel in the front level did not continue around the whole building. Rubble stone is more vulnerable than the fired brick, and thus requires more ring beams.*

4.2.3b This building has a heavy load-bearing façade on the front with a dhajji dewari *structure on the side. Timber lacing is not in evidence on the front façade.*

4.2.3 INTEGRITY OF THE SEISMIC BANDS: DO THE *TAQ* LADDER-BANDS OR DOES THE *DHAJJI DEWARI* FRAMEWORK ON ANY GIVEN FLOOR LEVEL EXTEND AROUND THE ENTIRE BUILDING?

CONDITION ASSESSMENT: *Taq*: It is important that the timber lacing of the *taq* construction be continuous around the building at each floor level. If this is not the case, the structure is at risk in an earthquake. The greatest benefit of the timber lacing in *taq* is that it is holds the corners of a building together. If absent, corners are likely to suffer the most damage from both earthquakes and differential settlement.

Dhajji dewari: It is important that the timber frame of *dhajji* construction is continuous around the entire building around the building at each floor level unless tied to the timber lacing in a *taq* wall. If this is not the case, the structure is at risk in an earthquake.

4.2.4 MASONRY PANEL STABILITY: *Dhajji Dewari*: ARE THE MASONRY PANELS LARGE OR SMALL? DO THEY EXTEND FROM FLOOR TO CEILING OR ARE THEY SUBDIVIDED? DO THEY CONTAIN DIAGONAL BRACES?

CONDITION ASSESSMENT: In *dhajji* construction, oversize floor-to-ceiling panels are vulnerable to falling out. Post-earthquake observations have confirmed that smaller panels on average show greater resistance against the loss of infill masonry. *Dhajji* walls in gables are particularly vulnerable because (1) the earthquake forces can be larger at the top of a building, and (2) the lack of overburden weight from walls results in a looser fitting masonry panel (see Figure 4.2.4b).

The question of whether diagonal braces are actually necessary is less clear, because in their absence the masonry panel (particularly brick masonry) is usually stiff enough to serve as a brace, and the difficulty of cutting and fitting the masonry around the diagonal brace leads to less tightly packed infilling. Diagonal braces were common historically and should always be left intact, but when they were not included in the original construction, this does not necessarily mean the structure is more vulnerable, especially if the masonry is well laid and the panels are small and tightly filled. An example of a well constructed *dhajji dewari* building without diagonal braces can be seen in Figure 4.3.1f.

In rural mountain areas where stone instead of brick is commonly used, the criteria for good performance is different. Traditional *dhajji* construction has evolved in rural mountain areas in a way that is responsive to the difference in the behaviour of the infill. Quite commonly, the buildings have much smaller infill panels – as is shown in Section 1.3. Photographs 1.3e to 1.3l taken in the rural mountain areas of Pakistan Administered Kashmir all show a smaller surface area of the walls than do photos 1.3a to 1.3d. There is also a greater

4.2.3c *This example in Bhaktapur, Nepal shows how the absence of complete bond beams can lead to the spreading apart of the masonry.*

Unlike in Kashmir, complete timber bond beams were less common in Nepalese traditional construction except in some of the palaces and temples. As a result many buildings collapsed in the 1934 earthquake. The Nepal building code now specifies the same ladder-style bands for masonry construction as are in Kashmiri *taq* construction and which are also in the current Indian Standard codes.

differentiation in the length and thickness of timbers. The quality of the stone used for the infill also varies greatly because it is dependent on what is locally available. For stability of the infill, the better stones are flat, or have been broken to produce two flat sides. When rounded stones are used without prior breaking or shaping, the infill can quickly become unstable in earthquakes, as can be seen in Figure 4.2.4d of a building damaged in the 20 February 2009 tremor, which was much smaller than the 2005 earthquake.

It is important that a condition assessment of *dhajji* construction with stone infill include an evaluation of the sizes and the shapes of the stones used for the infill masonry. It is also important to verify that the frame is complete, and reasonably well nailed or mortised together, and that the smaller elements, of the size of those shown in Figure 4.2.4d, are nailed to the surrounding framework at both ends.

REPAIR AND STRENGTHENING RECOMMENDATIONS:
The recommended maximum spacing of timber frame elements for infill masonry walls is 1 metre between studs (vertical members) and 1 metre between horizontal framing members (Arya & Agarwal, 2005). For the mountain rural stone infill type, this spacing is adequate only if this frame is further subdivided by the kind of smaller wooden subdivisions shown in the different types illustrated in Section 1.3. The masonry must be tightly packed into the frames. For existing oversize panels, it is recommended that they be reduced in size, rather than simply braced. Not only does this reduce the risk of falling masonry, it also improves the performance of the building by increasing

4.2.4a *The* dhajji *panels in this barn in Fatehgarh Village, west of Baramulla, fell out in the earthquake because they were large, and also because of the lack of overburden weight. The gable above had no wall prior to the earthquake, and the studs were also weakened from decay.*

4.2.4b *The masonry in this gable in a building in Baramulla was thrown out by the 2005 earthquake.*

the flexibility and energy dissipation of the structure, while ensuring against the loss of resistance when panels fall out of the frames.

Panels can be reduced in size by dismantling the top portion of the panel and introducing a timber crosspiece nailed with galvanized nails to the vertical studs. For gable walls, it is important to be sure that the masonry is secured into the frames at the top because these panels have no overburden weight.

While at a minimum, large panels can be secured from falling out of the frame by mounting a timber board across the top and at mid-height, this is not ideal. Surface mounted boards may deteriorate rapidly, and the oversize panels do not perform in an earthquake as well as ones that are within the recommended maximum size. External surface-mounted boards also disfigure the appearance of the building. For interior walls, however, particularly those near to beds, attaching timber trim along the top of the infill panels to keep the upper part of the infill from possibly falling onto people is an inexpensive and sound way of protecting against injuries. Since interior walls are usually plastered, the installation of a metal plaster lath can also protect against falling masonry. This protection against falling debris is particularly important for buildings with irregular stone infill. Cement plaster, however, should be avoided for the reasons stated in Section 4.1.3.

4.2.4c New post-earthquake dhajji *house being constructed in rural Pakistan Administered Kashmir showing the packing of the infill-frame with stones with clay mortar. This construction – with thin boards for studs and many small pockets for the masonry, is found in the rural mountain areas of Kashmir on both sides of the Line of Control where rubble stone, rather than fired brick, is used. Photo by Maggie Stephenson, UN-HABITAT.*

To avoid the problems shown in Figure 4.2.4d of *dhajji* with stone infill, the individual stones are most stable if they are laid with a flat surface in horizontal position. A good rule-of-thumb to observe is that stones in the wall should be placed as they would lie if individually placed on a table, with as large a flat contact area as possible. Their plane of contact with the next stone above should also be as flat as possible. In addition, all stones, except very small leveling stones, should extend through the thickness of the wall. All timber elements should be securely nailed to each other and the surrounding framework.

4.2.4d This building, damaged in the 20 February 2009 earthquake in Pakistan Administered Kashmir, demonstrates how the use of undressed rounded stones together with a poorly constructed and incomplete dhajji *frame can lead to collapse of the wall even in smaller tremors. Photo by UN-HABITAT.*

4.3 Evaluation and Repair of Structural Conditions Resulting from Alterations and Additions

4.3.1 Alterations with Structural Impact: Are interior walls timber-laced and solidly connected to the exterior walls, and is there an adequate number of walls? Have larger windows or door openings been cut into the exterior walls, or have exterior or interior walls been removed or reconstructed?

Condition Assessment: The earthquake resistance of traditional construction is a function of having sufficient numbers of walls at each level to carry the loads to the ground (see Section 2.2.1,

Load Path). Structural problems in earthquakes arise when houses are converted into shops, or larger windows and doors are installed, or when walls are removed to enlarge rooms. In fact, the reconfiguration of the interior of heritage structures in Srinagar, including removal of walls, is now a common trend. Any of these actions needs to be examined from an engineering perspective before being carried out, and new walls of the same system may need to be constructed to fill openings that are found to be excessive. There are two rules of thumb: (1) the walls on each floor in each direction need to follow a reasonably symmetrical pattern to avoid torsion in an earthquake, and (2) there needs to be roughly the same amount of wall area on each floor in multi-storey structures.

Strong timber window frames can also contribute to the resistance of a wall, as they can resist the spreading of the masonry into the window opening. However, if they are not well attached to the masonry, this benefit is diminished. In the example shown from Nepal (Figure 4.2.3c), the large ornamental carved window lintel penetrates the masonry instead of being secured to it, so the masonry wall is weakened.

In *taq* or *dhajji dewari* buildings that have large open spaces, large shop-front openings, and/or few interior partitions, the ground-floor level may be at risk of collapse. This condition is known in earthquake engineering as a "soft" or "weak" storey (see Section 2.3.4(1)). This occurs because of the lack of walls to brace the structure at the base where the forces are highest.

Careful analysis of the symmetry of panel sizes and the numbers of walls on opposite sides of a structure can help to reduce the danger of damage from torsion. Crosswalls are defined as walls designed to behave compatibly with the rest of the structure, to help hold it together and dissipate energy.

REPAIR AND STRENGTHENING RECOMMENDATIONS: If the existing walls in each direction on any floor level are less than 75% of those on the levels below and above, add new *dhajji dewari* walls. This problem is particularly likely to be found at the ground-floor level, and vulnerability is highest at this level because the overburden weight is greatest. These walls can be of *dhajji dewari* construction in either *taq* or *dhajji dewari* buildings. A double layer of masonry with thicker timbers in the frame for *dhajji* walls may be necessary if the loads are large or if previously existing walls have been removed to create large spaces. (This does not apply to attic spaces, although some crosswalls in the attic level are often needed to reduce the flexibility of the roof to prevent masonry in the gables from being thrown out. See Section 2.3.4: Weight and Stiffness Distribution.)

Concrete infill is a poor substitute for masonry in *dhajji* walls. They are only 10 cm thick and extremely stiff, but brittle, and thus prone to collapse, Any reinforcing in them is prone to rust because of the likely poor compaction of the concrete.

4.3.1a & b These show dhajji *type of timber frames with poured concrete instead of masonry infill. Often, as shown in 4.3.1b, this concrete infill is easily collapsed out of the frames in an earthquake. Photos by UN-HABITAT.*

ROUGHLY THE SAME AMOUNT OF WALL AREA NEEDS TO BE ON EACH FLOOR IN MULTI-STOREY STRUCTURES.

- Do not make changes to the structure which increase its stiffness. Never add new walls using a stiffer construction system, like bearing wall brick or concrete block, and avoid removing existing walls, particularly at the ground-floor level of a building. Greater rigidity attracts larger earthquake loads.

- Avoid the introduction of reinforced concrete shearwalls to strengthen an open shop-front level unless they are part of a coherent and sophisticated engineering design. The stiffness and strength of concrete shearwalls can make them incompatible with traditional construction systems because they may cause the traditional structure to deform around them.

4.3.1c This is the interior wall of a house in near Muzaffarabad, very near the epicentre where the shaking was strongest. There was considerable damage to the dhajji *infill panels, but the house remained standing and the family continued to live in the lower floor. The infill masonry was stone with mud mortar and plaster. There were fewer walls than is typical, so the walls suffered a larger amount of damage than other* dhajji *structures nearby.*

4.3.1d This hımış *house in Adapazari has enlarged windows which, particularly on the ground floor, left little wall area remaining to resist the effects of the 1999 earthquake, which inevitably resulted in the greatest damage around the largest windows. Nevertheless, the timber framed house did survive in a city where the tremors devastated rows upon rows of reinforced concrete buildings.*

4.3.1e This building in Srinagar has very little wall area on the ground-floor, and is thus vulnerable to collapse. In its present condition, it is probably held up in part by the building next door.

- Unreinforced masonry used to replace a portion of a *dhajji* wall and framework as in Figure 4.3.1f is dangerous because it is initially very stiff, but will then tend to collapse from diagonal tension "X" cracks or fall out from out-of-plane forces because it lacks the timber armature and attracts a disproportionate amount of earthquake loads. Poured concrete infill as shown in Figure 4.3.1a & b is also dangerous even if the framework is retained.

- To avoid interior hazards, interior walls in *taq* buildings should be either (a) *taq*, (b) *dhajji dewari*, or (c) 100% timber. In *dhajji dewari* buildings or parts of buildings, the interior walls should be either *dhajji* or 100% timber. If the walls are of *dhajji* construction, they can make a contribution to the seismic resistance of the entire building.

- All interior walls must be securely tied to the other walls, the floor, and the ceiling in order for them to contribute to the building's structural resistance to earthquakes.

4.3.1f The lower left section of the first-floor façade of this dhajji *building in Srinagar has been rebuilt in solid brick, eliminating the* dhajji *sub-frame. This can create a zone of stiffness in the structure, which can lead to the breaking away of that section of the façade, which then weakens the rest of the building.*

A NEAR COLLAPSE IN BARAMULLA

This house in the historic centre of Baramulla suffered severe damage in the 2005 earthquake because part of the masonry infill on one end of the building had been removed. This caused the heavier upper-floors to twist (see Section 2.3.4(2), Torsion), which then caused the remaining two panels of masonry in the end wall of first floor to fall inward. The owner, Mr. Zargar, is shown in Figure 5.7m. He was 90 years old and still living in the house when the photos were taken in April, 2007.

This story illustrates a number of important points. First, it shows that simple vernacular buildings can have a long life. The house is a valuable historic resource in desperate need of restoration from its current state of near collapse, and from the continued risk it presents to its owner who needs help to know what to do. Second, it shows the resilience of the *dhajji* construction against collapse even when the structure has been seriously weakened by removal of some of the infill walls.

Finally, it reinforces the importance of making sure that there are adequate numbers of walls in place with continuity to the ground, and that they are maintained in good condition.

4.3.1g Perhaps the oldest still extant house in Baramulla showing near collapse of the first-floor level from the 2005 earthquake.

4.3.2a (top left) Bathrooms are needed even in historic houses, but better ways exist for their integration into such structures. In this case, the new facility was placed as an addition right above the front door. Apart from the aesthetic collision it represents, this particular example could collapse onto the escaping occupants in the event of an earthquake because there is little lateral support for the heavy rigid box resting on the thin columns. This is an example of a potential soft-storey mechanism.

4.3.2b (top right) The front facade of this house has been plastered with cement plaster, which is destructive to the underlying brickwork. Perhaps more destructive, however, is the fact that it appears that the windows in this main wing of the house have been raised, thus destroying the integrity of the original architecture of the historic house.

4.3.2c (right) This house in Srinagar has suffered the kind of alteration that has afflicted many heritage structures. The property has most probably been subdivided between different members of a family, with one owner seeking to modernize their portion by rebuilding the facade and changing the floor levels. This not only destroys the integrity of the heritage architecture, but also substantially increases the seismic hazard by cutting and removing a portion of the timber frame, leaving both parts more vulnerable.

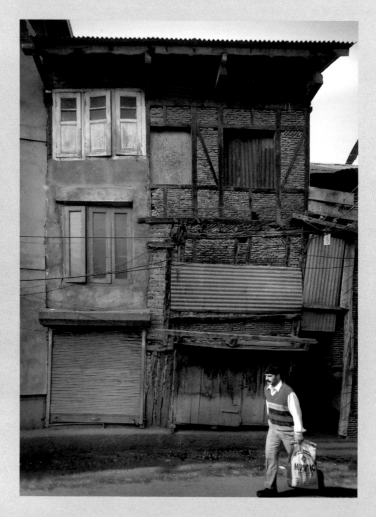

4.3.2 EXTERIOR ADDITIONS WITH STRUCTURAL IMPACT: ARE THERE MODERN ADDITIONS ON TOP OF OR BESIDE THE *TAQ* OR *DHAJJI DEWARI* BUILDING?

CONDITION ASSESSMENT: In recent years, roofs or upper-storey additions on many houses have been constructed in concrete block masonry or reinforced concrete on top of *taq* walls. This can be dangerous as the new parts are rigid blocks of great weight resting on the more flexible walls below. If there are building additions of different construction types on top of or next to the traditional building, these need to be checked for compatibility with the traditional *taq* and *dhajji dewari* building. This is particularly important for upper-floor additions.

Adjacent additions can also cause damage because the two parts with different stiffnesses can pound against each other with great force. Reinforced concrete with infill masonry and cement block construction can be extremely rigid and thus incompatible with the flexibility of traditional construction of any type.

REINFORCED CONCRETE WITH INFILL MASONRY AND CEMENT BLOCK CONSTRUCTION CAN BE EXTREMELY RIGID, AND THUS INCOMPATIBLE WITH THE FLEXIBILITY OF TRADITIONAL CONSTRUCTION OF ANY TYPE.

4.3.2d (above) A new bathroom with rigid concrete and tile walls had been installed in this hımış *house in Golcuk, Turkey. It was destroyed in the 1999 earthquake because it resisted the swaying of the building's more flexible timber and masonry structure.*

4.3.2e (left) View of a large taq *building in Srinagar with a top floor added later, probably in unreinforced concrete block masonry construction.*

Reinforced concrete slabs poured over the floors of either *taq* or *dhajji dewari* buildings are heavy and stiff, and thus potentially very dangerous. If such floors have been constructed, it is recommended that they be removed. In earthquakes in Italy, and also in the United States, stiff elements – even new concrete ring beams and concrete diaphragms installed for earthquake protection – have been observed to break out and be ejected from buildings during earthquakes, as seen in the case of a window that was filled in to create a shearwall in Figure 4.3.3a and a concrete roof structure that has contributed to the collapse of an upper storey in Figure 4.3.3b.

4.3.3a This image may be hard to discern as a chain link fence with barbed wire partly obscures the view. It shows, however, a surprising phenomenon. Prior to the 1994 Northridge, California, USA earthquake, this early 20th-century brick building had been retrofitted with reinforced concrete installed into window openings to strengthen the masonry wall. The historic masonry was laid with lime mortar, but the window infills were solid reinforced concrete tied to the masonry with grouted anchors. During the earthquake the forces ejected these rigid concrete window plugs from the wall by tearing the grouted anchors out. Had masonry with compatible mortar been used, it is likely that the wall would have absorbed the stresses more uniformly, resulting in less concentrated damage (see also Section 2.2.2 Strength and Stiffness).

4.3.3b Masonry house damaged in the 1997 Umbria-Marche earthquake in Italy. Rather than helping to hold the masonry structure together, the incompatibility of the heavy reinforced concrete roof with the masonry below helped to cause the collapse of the top floor of the structure. Photo by Sergio Lagomarsino.

Basic Structural Engineers' Checklist that applies to a range of building types (Based on *FEMA 310*)

1. **Check Load Path:** A load path must be capable of transferring the inertial forces generated by horizontal earthquake motion from the building mass to the foundation.

2. **Adjacent Buildings:** The possibility of structural damage or collapse from pounding of adjacent buildings of different floor levels and frequencies of vibration must be addressed.

3. **Weak storey:** It is recommended that the strength of the lateral-force-resisting system in any storey of a building should not be less than 80% of the strength in any adjacent storey above or below.

4. **Soft Storey:** It is recommended that the stiffness of the lateral-force-resisting system in any storey should not be less than 70% of the stiffness in an adjacent storey above or below, or less than 80% of the average stiffness of the three storeys above or below.

5. **Geometry:** It is recommended that any changes in the horizontal dimension of the lateral-force-resisting system (i.e. shearwalls, etc) be no more than 30% in a storey relative to adjacent storeys.

6. **Vertical discontinuities:** Vertical elements in the lateral-force-resisting system should be continuous to the foundation (with the exception of "crosswalls", that is interior *dhajji* or wood frame partitions.)

7. **Mass:** It is recommended that there be no change in effective mass of more than 50% from one storey to the next.

8. **Deterioration of Wood:** Check for any signs of decay, shrinkage, splitting, fire damage, or sagging in any of the wood members, and for deterioration, breakage or separation of any of the metal accessories, including nails.

9. **Masonry Units:** Make sure that all masonry units (bricks, adobe blocks, cement blocks or stones) are not cracked or deteriorated from salt or any other decay medium.

10. **Masonry Mortar Joints:** Check for the integrity and quality of the mortar to ascertain if it has maintained its original strength. In no cases should the mortar be powdery, and, in the case of lime mortar, it should resist being easily scraped away from the joint by hand with a metal tool.

11. **Cracks in Masonry:** Existing cracks in the masonry should be examined to determine the cause, particularly if concentrated in one location or if they extend beyond a timber band in *taq* construction. Any separation of the timbers in a timber band or frame in either *taq* or *dhajji* construction should be identified for repair.

12. **Cracks in Infill Walls:** The cause of cracks and gaps or offsets in the masonry panels in *dhajji* construction should be determined, and repairs made. The adequacy of the attachment of the panel to the frame should be analysed, and if found inadequate, strengthened.

For more general information on masonry and mortars see:

1. USA National Park Services Preservation Brief #2: *Repointing Mortar Joints in Historic Masonry Buildings,* by Robert C. Mack, FAIA and John P. Speweikon. (Available at www.nps.gov)

2. Heritage Society of British Columbia, *Masonry,* by Robert M. Patterson and P. Eng. (Available at www.vancouverheritagefoundation.org)

The following article includes reference to Kashmiri traditional construction:

1. "Bricks, Mortar and Earthquakes" by Randolph Langenbach, *APT Bulletin*, 1989. (Available at www.conservationtech.com)

Further information and complete structural checklists and repair and strengthening information on general building types (not Kashmir specific) can be found in *FEMA 310* and *FEMA 356,* which are available for downloading at www.traditional-is-modern.net "Library" page.

4.4 POST-EARTHQUAKE BUILDING INSPECTION AND EVALUATION

Most houses in an earthquake (those outside the epicentre) will be slightly to moderately damaged and there is a great potential for their reuse with traditional constructional and mending techniques. There is a tendency for the damage to be assessed as more severe than is the real case because damage assessment is by people without past and relevant experience.... Traditional building types have developed over long periods and are often sensitive to the location and climate.

Richard Hughes, Ove Arup and Partners, London, 2005[13]

The 2005 earthquake highlighted the need for rational and accurate assessments of buildings following a disaster. This may sound like a straightforward process, but where premodern construction is involved it is not. Many people who undertake damage assessments after an earthquake are both inadequately trained and unfamiliar with traditional construction methods, and thus often unable to distinguish between the kind of damage that puts people at risk in future aftershocks, and that which does not. In addition, they may not be able to judge what remedial measures one may take to protect a damaged building from further damage or collapse, or ultimately to repair it. They may thus act over-cautiously and recommend the abandonment or rapid demolition of structures that could be saved, even though repair would be far less expensive and disruptive then any new construction. Similarly, local inhabitants often undervalue traditional construction and idealize "modern" materials, especially when they are fearful and see damage like that in Figures 4.4c & d.

National and international programmes frequently aggravate this situation by providing generous amounts of funding for the owners of structures which their assessment teams determine to be "destroyed", while providing far less funding to those whose structures are determined to be reparable. This can encourage owners to plead with inspectors to condemn their houses so they can receive the replacement money. This situation leads not only to the destruction of structurally viable traditional houses, which compounds relief needs for the needlessly displaced families, but also precipitates the deterioration of the traditional built environment, the erosion of the local cultural identity, disruption to the community, and loss of local craft skills.

Efforts to help disaster victims will fail if they are unresponsive to the culture of the displaced people and to the traditional functioning of the community. This kind of failure has been observed in situations where displaced people are not actively involved in the repair of their houses that were not destroyed, or in the design and construction of new houses to replace those that were damaged beyond repair.[14]

4.4a *This family in Topi, a rural village near Bagh, Pakistan Administered Kashmir were living in a tent until they could finish reconstruction of their home on their own property nearby. The house visible behind (also in Figure 5.5e), of* dhajji *construction, survived the earthquake. (More about this village in Section 5.5.)*

4.4b *New cement-block houses next to the Kutch village of Vondh are part of a large resettlement area of cement-block houses for people displaced by the 2001 Gujarat earthquake. The white structures are the toilets. This "new village" is located on a bug-infested swamp and laid out in rigid rows that contrast sharply with traditional settlement patterns. The houses have few windows and no private yards or courtyards. Most of the houses were still unoccupied when this photograph was taken in 2006. This stands in sharp contrast to much of the reconstruction in Gujarat, where some 80% of it was owner-built, and did not involve such massive settlement relocation.*

4.4c *A version of stud frame with masonry infill construction is known as "bahareque" in Central and South America. As as long as the timbers are not rotted, this system resists collapse in earthquakes. However, because of their flexibility, the plaster falls off, which to inspectors lacking knowledge of the system, can lead to unnecessary condemnation. Almost all of the plaster was shed from the walls of this multi-family house in the 1986 San Salvador earthquake, but there was no structural damage to the underlying walls.*

The distressing examples of the failure of post-earthquake relief efforts stem from the initial mistaken premise that the typical plaster and masonry damage found in earthquake-damaged traditional buildings means that the buildings cannot be safely repaired because their construction systems are old and weak. The damage assessment of traditional buildings must thus be improved in order to recognize the different behaviour and potential for reserve capacity of traditional buildings versus contemporary structures.

The essence comes down to one fundamental difference between concrete buildings and timber-laced traditional structures. As mentioned, traditional structures have proven to be robust in earthquakes from their combination of flexibility, energy dissipation and redundancy. While they manifest damage (cracked plaster and loose or fallen bricks) easily at the onset of shaking, this level of damage is not determinative of a life safety risk. This was demonstrated by the tests of the Portuguese Pombalino walls and the stability of Turkish *hımış* houses in the 1999 earthquake described in the previous chapter. By contrast, many reinforced concrete with masonry infill structures collapse because of the rapid degradation of their beam/column joints. They lack redundancy because their walls are not designed to be part of the structural system, so mistakes in construction of the structural frame can be catastrophic.

In an earthquake, a highly damped flexible response can save a structure from collapse, but the stiff elements within it will nonetheless be cracked or shattered as the structure moves. Exterior and interior plaster is made with weak but inflexible materials. This is especially the case with modern cement plasters but it is also the case with mud and lime plasters. Thus, any earthquake will almost immediately cause a traditional structure to shed its plaster into heaps of rubble on the floor

TWO EARTHQUAKES: A COMPARISON

4.4d & e *The top photo was taken in Adapazari one month after the great 1999 Izmit Earthquake in Turkey (magnitude 7.6, fatalities: 17,000), and the bottom photo was taken in Orta one day after the 2000 Orta earthquake (magnitude 6.0, fatalities: 2). Not surprisingly, the top one was more damaged, but when one takes into account that the Adapazari house had long been abandoned, the similarity in damage demonstrates the resistance of* hımış *construction to collapse in large, as well as small, earthquakes. Almost all the fatalities were in modern concrete buildings.*

4.4f (above) & 4.4g (right) Typical house and landscape in Yuva. 4.4h (far right) New houses in Yuva

A Tale of Two Villages:
Yuva and Elden, Turkey after the 2000 Orta Earthquake

Two rural farming villages near Orta, Turkey, Yuva (above and under this text), and Elden (below), suffered damaged in the 2000 earthquake. Rather than helping the residents repair their houses, the government recommended the villages be relocated to what was described by their geologists as safer ground. The residents voted to accept the government's proposal, which came with the promise of new houses in exchange for their old ones. In September 2004, more than 4 years after the quake, the new construction was still not finished, and only a handful of the houses had been occupied. The new sites were selected by a geologist, but not by an agricultural expert or social scientist. They were remote from all that is otherwise necessary for human agricultural settlements – water, trees, fertile soil, and protection from the wind. No provision was made for barns for the animals or for a community centre, general store, or even a mosque. In the case of Elden, the new village was many kilometres from the old. The new houses were constructed by government contractors like barracks in rigid rows, and, ironically, many are of unreinforced masonry (Figure 4.4h), and thus less safe than the ones they replace. In Elden, five years after the quake, fewer than ten of the 80 new houses had ever been occupied, and several of those had been abandoned when their occupants returned to the old village to be near their fields, animals, friends, and mosque. Most in both villages, including the elders in Figure 4.4j below, now regret their earlier decision to accept the relocation.

4.4i (above) Elden Village. 4.4j (right) Men of Elden outside the mosque. 4.4k (below) Elden New Village

4.4l & m These children in Baramulla are pointing to the cracks in the plaster caused by the earthquake, which are barely visible in this photograph. In spite of the seemingly small amount of damage to this house overall, the family had mostly moved out after the earthquake because of their fear for the safety of their taq *home, seen in the top photo.*

or on the ground. While most people in a calm moment may realize that the plaster surfaces on a building are cosmetic and not structural, in the post-earthquake environment it is often hard for people to comprehend this when they see many large cracks and the debris from partially collapsed plaster wall surfaces. Their homes seem to them, and also to the surveyors and engineers doing the assessments, to be heavily damaged, even though they may in fact not be.

This confusion can be further accentuated by the fact that if the buildings had been built of reinforced concrete, the same level of plaster damage most probably *would* indicate a serious safety concern. Thus, the failure to distinguish between structural and cosmetic damage can rapidly lead to the buildings' abandonment, condemnation and eventual demolition.

For these reasons, it is critically important to emphasize the fact that damage in the form of cracked plaster, or even dislodged infill masonry, should not be taken as evidence that a traditional building is unsafe and must be evacuated. Both the *taq* and the *dhajji dewari* systems are resilient. They resist serious damage and collapse in earthquakes because of the friction between the masonry and the wood, and within the masonry itself. It is this controlled level of damage that prevents collapse. In almost all cases, these timber-laced masonry buildings can be safely repaired and, once repaired, the building will be as safe and stable as before the earthquake. The fact that they are reparable is one of the benefits of this kind of construction when compared to reinforced concrete. Once the concrete frames are significantly cracked or deformed, reinforced concrete buildings are difficult or impossible to repair to their original strength.

4.4n & o In contrast with Yuva and Eldon, most of the members of the nearby farming village of Asagi-Kayi did not seek government assistance for new houses, and the government did not propose to relocate the settlement. Instead, they proceeded to repair their houses in less than a year after the earthquake using traditional methods, and went on with their lives. The photo on the left was taken one day after the earthquake, and the one on the right shows the same room after the repairs.

Leepa and Neelam Valleys in Kashmir: An Example of Corrective Measures Taken in Time

Maggie Stephenson of UN-HABITAT in Pakistan reported in February, 2009: "The Leepa and Neelum valleys are remote areas with long and continuing traditions in multi-storey timber frame construction. They suffered damage but not heavy destruction in 2005 despite proximity to the epicentre. Because they were timber and stone houses, they were first categorized as *kacha* (informal), and advised to rebuild in 'safe' new masonry construction. Many people started accordingly and made new plinths with steel in order to collect their next ERRA payments.

"The community meanwhile argued that their traditional construction had performed well in the earthquake and met their needs and preferences in areas with 8 feet of winter snow, extended families, and high levels of local skills in stone and timber work. UN-HABITAT documented the buildings and their performance, and advocated that they represent not only important architectural heritage but also indigenous engineering and environmental knowledge. These buildings represent a living sophisticated building tradition and repository of skills. Houses are usually constructed with a stone base and with several storeys above constructed in *dhajji* or with timber cavity walls. Elaborate systems of heating, verandahs, wood carving and decoration add further to their use and cultural value."

"On the advice of NESPAK and UN-HABITAT, ERRA agreed not only to endorse 'Leepa' type construction as eligible for financial assistance, but also agreed to grant the same level of financial assistance for appropriate repair of existing traditional houses as for construction of new ones. This policy has not only safeguarded the future of an important stock of buildings, but also generated large investment in conservation, demand for these skills, and reduced the environmental impact of the reconstruction. The ERRA endorsement and training of carpenters has also increased the use of traditional construction for new buildings in preference to reinforced concrete. Today you can see abandoned concrete plinths, where the people took up the opportunity to repair or construct their homes instead ,according to the training provided by UN-HABITAT based on local best practice in close consultation with the local master carpenters."

4.4p (Top) Family on timber frame varandah of traditional Leepa Valley house after earthquake. 4.4q Five and a half-storey house with typical elaborate multi-storey verandahs. 4.4r Now abandoned concrete plinth constructed after earthquake before rule change that allows for restoration of damaged wooden house. Photos by UN-HABITAT.

4.4s *After the 2005 Kashmir earthquake, this family near Uri were continuing to live in their damaged home, but were fearful, especially since government inspectors had told them their house was unsafe. The house, with a combination of* taq *and* dhajji *construction, had many cracks, but its structural integrity was intact and, contrary to the inspectors' statements, it easily could be repaired. The family would be best served if they received appropriate advice on how to repair their traditional house and financial assistance for that purpose. (A view of the interior of this house showing the characteristic earthquake damage can be seen in Figure 3.4c.)*

POST-EARTHQUAKE ASSESSMENT RECOMMENDATIONS IN A NUTSHELL

1. Inspectors should be trained on the structural attributes of traditional construction. Much of the current training of engineers has little to do with traditional construction, and the socialization of many engineers into the profession is directed at cultivating the belief that traditional construction is structurally unsound, antiquated and obsolete.

2. It is important that inspectors receive sufficient training in heritage conservation to be able to identify historic vernacular buildings and to understand the differences in treatment such structures should receive during the post-disaster reconstruction period. Important historic resources are often destroyed after disasters because of uninformed official or private actions, even though they had survived the disaster itself.

3. It should not be assumed that traditional buildings with lots of cracks and fallen plaster cannot be repaired, nor should it be assumed that damaged masonry, timber-laced masonry walls, or *dhajji* infill walls cannot be repaired back to equal or better than their original structural capacity. (By contrast, damage to reinforced concrete buildings is often difficult to repair to original strength.)

4. It is recommended that inspectors separate falling debris hazard assessments from risk of building collapse. Falling hazards can be addressed quickly – often by owners themselves. Restricting access or occupancy for

this reason, without distinguishing it from the risk of collapse, can easily lead to the permanent displacement of the owners and destruction of the dwelling.

5. People are fearful after an earthquake, so it is important that experts who are called upon to inspect their houses do not aggravate those fears unnecessarily. They need to work with the occupants to give them a clear understanding of what needs to be done to repair their dwellings sufficiently for occupancy, rather than encourage replacement.

6. Government financial assistance in a number of different countries often is structured so that a premium is provided to owners whose houses are deemed to have been damaged beyond a certain level by inspectors. Government should be discouraged from making this premium so large as to encourage the condemnation of buildings that can be repaired. In the past, this has created a greater refugee problem and it adds enormously to the publically funded cost of the disaster, as both the temporary and the new permanent housing is far more expensive than expeditiously done repairs. The replacement, rather than repair, of houses can lead to the destruction of communities.

5a This forlorn fragment is all that is left of a *dhajji dewari* house that was cut exactly at the subdivided property line disregarding its tilt when half of its site was redeveloped with a large reinforced concrete dwelling.

A·S·B

TAQ AND *DHAJJI DEWARI* VS. REINFORCED CONCRETE:
TRADITIONAL IS MODERN

You have to have some sense of cohesion to urban form. Too much of what we have built in India today looks like the bottom of the sea. It is just everything plunked down wherever it fell.

Charles Correa, quoted in the
New York Times, 2007 (Kahn, 2007)

Much of the world has gone through a veritable revolution in building construction methods over the course of the last half-century. Traditional ways of building have been displaced in countries across every continent, and the principal system that has supplanted them has been reinforced concrete – particularly reinforced concrete moment frames with masonry infill. Reinforced concrete is often viewed as strong and "modern", while traditional construction, including *taq* and *dhajji dewari*, is seen as obsolete, insubstantial, and symbolic of poverty.

A large part of this misconception comes from constant pressure from the construction industry, commercial interests, and agencies of the government, including some international intergovernmental agencies and NGOs, which work hard to convince people that such traditional buildings are unsafe as well as obsolete. Safety and modernity, they say, can only be achieved with new construction in reinforced concrete. One international consultant, a conservator, experienced this kind of problem in Afghanistan at an international donors' meeting for post-war reconstruction work. He described how one intergovernmental agency official asked: "How many cement plants do you need?" Then, in response to advocates for using traditional mud brick construction, he said, "it is disgusting, unsanitary…the stuff of peasants."[15]

As a result of these views, the architectural heritage in Srinagar and many of the other cities and villages in Kashmir is rapidly being lost. This has happened to a great extent not because the buildings have ceased to be serviceable, but because they have not been considered to be important. The examples of new reinforced concrete buildings in Srinagar show the profound impacts on the city of this kind of redevelopment. There is a change in the relationship of the buildings to the street and to the neighbouring buildings, and an incompatibility of scale and of the character of their surfaces, fenestration and other architectural details (Figure 5b).

5b The "camel's nose under the tent." This photo was taken in the 1980s in Srinagar after the construction of a new three-storey reinforced concrete building in the centre of the historic city, where it looms over its older three-storey neighbours. At that time, this kind of blockbuster building in the historic centre of Srinagar was rare, but a decade and a half later it sadly has become common.

It is easy to overlook the importance of each individual building until a large number of them have been replaced or disfigured by generic "modern" houses. Yet it is the way that Kashmir's traditional residential buildings relate to each other as a group that contributes to the area's magnificent natural and cultural landscape by giving it a unique historical character. Like the Sufi story of the man who disfigured himself to achieve enlightenment, it seems that it is only after witnessing such destruction that people later discover the worth of historic artifacts and their own need for them; but by then it is often too late to save more than a few remaining isolated examples.

5.1 THE IMPORTANCE OF THE "CULTURAL LANDSCAPE" IN LIVING CITIES

In Srinagar, the traditional vernacular architecture which embodies the cultural value of an extraordinary historic city has suffered greatly in recent years. The architectural language that gives traditional buildings their character and compatibility with each other is a language of cultural consensus. An effort to restore this can also contribute to a restoration of shared community values in the present day.

To illustrate this premise, it is worth looking at the view of the Old City area of Baramulla from the historic wooden bridge looking west towards the mountains (Figure 5.1a). Like a flock of birds on a sandbar in the river, the houses of this old section of Baramulla appear seamless with the magnificent mountains behind. They have become an essential part of this view – what UNESCO's World Heritage Centre would properly characterize as a "cultural landscape".

Now if one looks at the other side of the river, even from the same vantage point of the wooden bridge, one can see empty lots and new buildings, including one large brick and reinforced concrete building under construction (Figure 5.1b). If one now imagines that the empty lots and new buildings transposed onto the view of the old city seen in Figure 5.1a, it is clear that the overall scene would be irrevocably harmed. With this example, one can begin to see how the whole is more than the sum of its parts. One new building, or indeed one thoughtless demolition, can be the ruin of a precious ensemble.

In Srinagar, the view looking west from 3rd Bridge similarly is an iconic view, as can be seen by the photographs which were taken a century and a half apart and are reproduced in the Introduction. Many houses have been changed in subtle ways, but for the most part in a manner consistent with the heritage typology that had been established even centuries earlier.

For these and other views to survive, they must now be consciously protected, because in the present century Srinagar is poised on the threshold of what is potentially a radical and irreversibly destructive change. The effects of this can already be seen along other parts of the banks of the River Jhelum where the wooden bridges, like the one

5.1a Left: The old city of Baramulla from the wooden bridge over the Jhelum looking towards the mountains, 2005.
5.1b Right: view of the South Bank of the River Jhelum viewed from the old city, showing empty lots and new construction.

photographed in 1979 (Figure 5.1c), have almost all been replaced with concrete bridges more inspired by freeway viaducts than urban amenities (Figure 5.1d). This kind of change is even more profound when the waterways themselves are destroyed, as was the ancient Mar Canal (*Nalla Mar*), as is illustrated in Figures 5.7a and 5.7b. This formerly magical water passageway was converted to an ugly two-lane road lined with bunker-like concrete shops in the 1970s.

In addition to the demolition of heritage buildings, the historic integrity of Srinagar also is suffering from the remodelling of houses. In this case, ironically, the effort is often only directed towards the effacement of the original architecture and the substitution of a new "modern" look, as can be seen in Figures 5.2a and 5.2b, rather than providing any significant improvements to the quality of life of the occupants. Changes in internal living patterns can have a devastating impact on the historic character of what would otherwise have been well preserved houses, especially when those changes, such as the raising of windows, lead to the destruction of the exterior architecture of the buildings.

Ways of living do indeed change over time, but heritage buildings need not always be disfigured or destroyed to accommodate these changes. If left in their older configurations, such buildings serve to teach each succeeding generation something about life in the past, giving a richness to life in the present that transcends temporal notions of what may be "modern".

While it is beyond the scope of this book to explain in detail how older vernacular houses can be adapted for modern living, it is important to emphasize that there is nothing inherent in their construction that should be seen as preventing the installation of modern plumbing, electricity, bathrooms and kitchens. It takes creativity to do it well but, as millions of people living in heritage structures in every country on every continent have found, it is worth it.[16] Imagine Venice, for example, without its historic buildings, including their historic interiors. Now, in the 21st century, almost every residence in Venice has modern facilities. Srinagar, once affectionately called the "Venice of the Himalayas" because of its network of canals, can benefit from this lesson.

In addition to the case made for preservation, this chapter will explore the issue of whether either *dhajji dewari* or *taq* construction technologies can take their place alongside construction in reinforced concrete for new buildings as well as in renewed and preserved heritage structures. It is hoped that this discussion can overturn the myth that these buildings from an earlier time and technology are obsolete and unsafe just because they are not constructed with modern materials.

The following section discusses the problems presented by the almost universal adoption of reinforced concrete construction as the standard for all types of buildings in Kashmir, and the impact that

5.1c Top: 1979 View of the River Jhelum in Srinagar showing one of the old wooden bridges that have now been replaced. Photo by Tom Dolan.

5.1d Bottom: A view of the new 1st Bridge. It is as functional as a freeway viaduct, but not a thing of beauty or an asset to the urban environment.

5.1e G.M. Bhat, Professor of Geology, is here seen looking back into his own past by opening his family house in Pampore, outside of Srinagar.

5.1f A view of the kitchen in the Bhat house, which remains exactly as his father left it when he died in 1992. Within a single generation, this kind of interior and the lifestyle of which it was a part have gone from being commonplace to almost unique, especially in the cities. Interiors like this are often deliberately expunged when modernized. Now, when found in such a state of suspended animation, the effect is even more striking than if it were an exhibit in a museum.

this has on the historic urban fabric of Srinagar and other cities in Kashmir. The second section documents examples of new buildings using traditional construction techniques in Indian and Pakistan Administered Kashmir. The third section discusses sustainability and "green" architecture, and how traditional masonry and timber construction should be considered to be compatible with these goals, even for new buildings.

5.2 Reinforced Concrete and the Pursuit of the "Modern"

Concrete is not a new material: it has been around for centuries. The dome of the Pantheon in Rome, Italy, one of the largest and most famous domes ever constructed, survives today with a structural capacity little diminished over the almost 2,000 years of its existence (Figure 5.2c). Ironically, had the dome been steel-reinforced like a modern reinforced concrete building, its longevity probably would have been less than a century due to corrosion of the steel.

Steel-reinforced concrete was introduced in the 19th century. It is not a material, but rather a highly engineered construction system. This is its strength – but also its weakness. If used with a high level of expertise and craftsmanship, reinforced concrete is extraordinarily robust, even in earthquakes. Examples in Japan, the west coast of the USA, Italy, the Middle East, and also in India have proven this fact. However, without this level of sophistication in design, material quality, and construction, and without the consistent use of proper construction equipment (particularly, for example, a mechanical vibrator, and even a good sized mechanical cement mixer, as will be explained in Section 5.3), reinforced concrete construction can be a catastrophe waiting to happen, sometimes even with relatively small tremors.

This is especially true for the moment frame with masonry infill wall design that has become the standard system in Kashmir, as it has in most other earthquake areas in the world. To be reliably safe during large earthquakes, reinforced concrete buildings (moment frame structures in particular) must conform to satisfactory standards in (1) engineering design, (2) quality of materials, and (3) quality of construction. The vast majority of non-engineered reinforced concrete moment frame structures in many earthquake-vulnerable areas of the world are seriously deficient in at least one, if not all three, of these categories. Many of these deficient structures have already cost tens of thousands of people their lives.

By comparison with moment frames, reinforced concrete shearwall buildings and confined masonry buildings have demonstrated a much better track record, but for a number of reasons they are less common in Kashmir and the rest of India and Pakistan. In fact it is surprising

5.2a & 5.2b A new fake brick cement plaster layer has been applied to the surface of the historic masonry of this traditional taq *house in Srinagar in an effort to make it look "modern."*

In addition to the destruction of the cultural value of the building, this kind of change can lead to the rapid deterioration of the underlying masonry and timber by trapping moisture under an impermeable layer of cement. In a future earthquake, the flexibility of the underlying building can cause this added layer to crack and fall off, with potentially lethal consequences for anyone attempting to leave the building during the shaking.

5.2c The interior of the Pantheon in Rome. The Pantheon was completed in AD125 during the reign of Emperor Hadrian, and its 142 foot (43 metre) diameter dome of unreinforced concrete was not surpassed in size for over 1,300 years.

5.3a The absence of shearwalls places this reinforced concrete building, seen under construction in Mingaora, Northern Pakistan, and the thousands of others like it, at severe risk of collapse in future earthquakes.

to see how uncommon confined masonry is, considering both its ease of construction and robustness. Confined masonry is a reinforced concrete frame construction system where the masonry for each floor is constructed first, and then the concrete frame is poured around it. It can be more economical to build because formwork is needed for only two sides of the columns and only the sides of the beams that border the infill masonry. This system has been found to be both more forgiving of compromises in the quality of design and construction than standard infill moment frames and is certainly better than unreinforced masonry. (For illustration, see Appendix 2: Glossary. For information on confined masonry, see www.traditional-is-modern.net.)

Disaster specialists have recently become aware that problems with reinforced concrete infill frames are more than a minor concern for some remote corners of the developing world. At the 13th World Conference on Earthquake Engineering in August 2004, for example, Fouad Bendimerad, Director of the Earthquakes and Megacities Initiative (www.emi-megacities.org), reported that "approximately 80% of the people at risk of death or injury in earthquakes in the world today are the occupants of reinforced concrete frame infill-masonry buildings." Poor design and bad construction are indeed a good explanation of the collapse of many concrete buildings. With the intense development pressures faced in developing countries, it is unlikely that this problem will be overcome in the near future. The problem of earthquake hazard reduction is not primarily an engineering problem. It is a socio-economic problem. Poorly constructed buildings will inevitably outnumber the better ones for a long time to come.

5.3 Reinforced Concrete: Its Promise and Shortcomings

Concrete construction requires more than just good craftsmanship: it demands an understanding of the science of the material itself. The problem is that builders are often inadequately trained and thus do not fully understand the seismic implications of construction faults. As a result, potential disaster lies hidden beneath the plaster. Invariably it is discovered after earthquakes that workmen have troweled over construction faults, including rock pockets, exposed rebars, and inadequately hydrated cement, all of which characterize construction done without the necessary knowledge and equipment.

Traditional buildings, even the ones that have survived earthquakes that felled nearby reinforced concrete buildings, were not engineered. No calculations were made, and no plans for them were ever inspected because none were ever drawn. They were constructed by local masons with little or no formal training and without the input of professionally trained engineers or building designers. They were built with a minimum of tools, with locally acquired materials using a minimum of costly resources, and they were held together with a minimum number of nails and fasteners. In many, the timber was not milled, being only cut and de-barked. Their frames were sometimes nailed together with only a single nail at the joint before the interstitial spaces were filled with brick or rubble stone in clay or weak lime mortar. The resulting quality of traditional buildings varies greatly, yet those with the kinds of timber-laced structural systems described in this book have repeatedly proved able to survive large earthquakes.

From its inception in the early part of the twentieth century, reinforced concrete has held great promise. It is manufactured from a widely available resource (limestone, sand and gravel) and it can be mixed and poured into any shape conceivable. It has captivated engineers and architects alike because of its capabilities, yet in earthquakes its record has been disappointing largely because of the pervasive quality control problems endemic to the material.

Unlike traditional timber and masonry, reinforced concrete requires a high level of knowledge and skill even to meet the basic levels of capacity and ductility to ensure against collapse. For example, the failure to use a mechanical vibrator to ensure proper flow and consolidation of the mix when pouring a structural column or beam can almost guarantee potentially devastating faults. The hand mixing and placement of the mix without vibration (Figure 5.3e), rather than directly from a transit cement mixer (Figure 4.1.4a) introduces gaps and cold joints into critical structural elements (see Figure 5.3f and also Figure 4.1.4b). Careless mixing of the ingredients, without precise measuring, can dramatically reduce the strength of the element. Even when all these steps and others are done perfectly, if the resulting concrete elements are not kept wet for 30 days, the hardening process cannot progress to design strength. All of this takes training and knowledge, both of which are in short supply in many

5.3b Collapsed residential blocks in Gölcük, Turkey, after the 1999 Kocaeli earthquake. Row upon row of these blocks suffered partial or total pancake collapse.

5.3c Srinagar and the rest of Kashmir are festooned with advertisements for cement, here shown in front of a new commercial building out of scale and character with its surroundings, just a short distance from the Jama Masjid.

5.3d The "Khyber" sign shows the imaginary modern city.

5.3e Concrete mixing for a new house to replace one destroyed in the 2002 Afyon earthquake in Turkey. The mixing with a garden hose and a hoe directly on the ground lacks the cleanliness and precision necessary to ensure adequate strength.

5.3f An extreme example of concrete placed without the use of a vibrator in a Turkish mid-rise apartment building under construction at the time of the 1999 Kocaeli earthquake. The building was damaged beyond repair by the earthquake before it was completed.

rapidly developing regions of the world where concrete has become the principal material of choice.

Reinforced concrete frame construction has been introduced into a premodern building delivery process without changing the underlying limitations of that system. The casual rural system of local builders with only a rudimentary knowledge of the science of materials had been sufficient for timber and masonry. However, with the introduction of concrete, it has proved to be woefully inadequate. Since reinforced concrete became the default choice for almost all new residential and commercial construction, this problem has expanded exponentially. The modern buildings possess the same level of deficiencies in construction quality as the traditional buildings. The traditional buildings have often remained standing in large earthquakes, however, whereas many of the concrete buildings collapsed.

It appears now that we have one system constructed with strong materials that is subject to catastrophic failure in large seismic events if it deviates even in small ways from a highly sophisticated level of perfection in design and construction; and another considerably less sophisticated system constructed of weak materials by relatively untrained craftsmen that is with few exceptions robust enough to withstand major earthquakes.

With the rapid spread of reinforced concrete construction during the middle of the last century, traditional construction has been displaced from all but the most remote rural regions in many countries within a single generation. This represents a transformation of the building process from an indigenous one to one more dependent on outside contractors, specialists, and nationally based materials producers and suppliers of cement and extruded fired brick and hollow clay tile.

In India, from 1981 to 2003 cement production rose from 20 million metric tons per annum to 110 million metric tons. As of 2003, the capacity of the plants was 140 million tons per annum and rising.[17] In Kashmir, one need not travel far to notice the proliferation of signs advertising cements from the many companies now competing for business (Figures 5.3c and 5.3d). Reinforced concrete has for all intents and purposes taken over what was once a local owner-driven building industry. It is analogous to a situation that would exist if all restaurants and markets were to sell hamburgers instead of preparing the rich local Indian and Kashmiri cuisine or selling locally grown produce.

Nevertheless, this description of the problems with reinforced concrete should not be taken as advocacy for its discontinued use. Concrete is here to stay, and it would be unrealistic to propose otherwise. As energy and other resources have ceased to be cheap and plentiful, it is important to consider making reinforced concrete construction take its rightful place alongside other less energy-consumptive materials and systems, rather than be used for almost every urban structure.

5.4a This man lived with his mother in an unreinforced stone masonry house in Bhuj, the ruins of which are behind him, when the 2001 Republic Day Gujarat earthquake struck.

After two weeks in the hospital from injuries sustained, he returned to the site to dig up the body of his mother, which he was cremating in the fire visible behind him. This photograph was taken one month after the earthquake. He is holding a photo and identity card of his mother in his hands.

Had there been a timber-laced construction tradition in Bhuj, as there had been in Ahmedabad, his home might not have collapsed.

5.4 UNREINFORCED MASONRY WITHOUT TIMBER LACING

The traditional timber-laced construction methods have not only been lost in favour of reinforced concrete, but, in urban areas, have also increasingly been ignored in favour of unreinforced masonry. Beginning in the days of the British Raj, the kind of unreinforced masonry construction common in the rest of India became standard in Kashmir, as it was, for example in the Uri Cantonment. With their dressed stone facades without timber-lacing, unreinforced masonry buildings were clearly seen as more permanent and prestigious, but the performance of unreinforced masonry in the 2005 earthquake was particularly poor. In Uri, for example, all of the school buildings and the mosque in the Cantonment, which were of unreinforced masonry construction, collapsed. Many other masonry buildings were damaged beyond repair (see Figures 2a, 3.1f and 5.4b).

While reinforced concrete, when correctly used, promises good resistance to earthquakes, the 2005 earthquake did show that masonry without some kind of reinforcement, even if seeming to be well constructed, was even more prone to collapse than poor quality reinforced concrete buildings. The problem was aggravated by the frequent use of rubble masonry behind a layer of dressed stone, the use of dressed stones that are almost square, and the almost complete absence of bond stones in stone work or bond courses in brickwork to tie the inner and outer layers of masonry together. The frequent use of rubble masonry in Kashmir proved particularly lethal. This problem extended through the damage district in both India and Pakistan.

5.4b This middle school building in Uri illustrates a common problem with masonry in India and Pakistan. The stones are too short for a proper bond. Many are rounded river stones. There is no evidence of through-wall bond stones. All the buildings in this large school complex collapsed, most of them completely. In Pakistan, over 6,000 schools were destroyed. UN-HABITAT reported that most rural schools there suffered from similar problems. (See also Figure 2a)

5.5a Wooden 18th- and 19th-century houses in Marblehead, Massachusetts, USA. These houses are part of a listed historic district, almost all of them are privately owned homes, not museum houses, yet heritage conservation is embraced by the entire community and is now essential to the economy of the town and the region. All of the houses in this view are 100% constructed and clad with timber.

5.5 NEW *TAQ* AND *DHAJJI DEWARI* BUILDINGS?

Why would anyone wish to build a timber and masonry building when they could have (maybe even for the same or less money) a modern house in reinforced concrete? While heritage buildings may be worth preserving, the idea of using archaic systems for modern construction may seem inappropriate to many people in Kashmir, and in many places of the world.

Ironically, in developed countries, the use of traditional methods and systems has in the last 30 years become popular. A real revival of traditional techniques mixed with the most cutting-edge scientific achievements has over the last two decades yielded some very interesting "green" buildings. In Germany, for example, the German Association for Building with Earth produces a giant trade fair and conference every four years. In France, a school and research institute on earthen construction, CRATerre-ENSAG, researches and promotes the construction of earthen buildings in Europe and the Middle East and has been involved in the development of new houses in France constructed entirely of unfired clay and timber.

In the United States and Canada, over 95% of the population live in houses constructed almost entirely of wood. Until the

5.5b This new building of dhajji dewari *was photographed during the 1980s while under construction in the centre of Srinagar.*

mid-19th-century, heavy timber braced frame construction was used. Since sawn stud "balloon frame" construction was invented in the 1830's, North American timber framing for new construction has changed little over the last century and a half, except that plywood has now replaced diagonal sheathing (see Section 1.2 for more information, and page 132 for a photograph of typical North American stud frame construction).

In the American Southwest, adobe and straw bale construction has become increasingly popular, both for aesthetic reasons and because of its superior thermal attributes. In Central America, timber and masonry *bahareque* construction has never ceased to be popular. In Peru, mud and timber *quincha* construction is still practised, carrying on a tradition that, as mentioned in Chapter 1, has lasted for at least 5,000 years. Examples of such innovative use of traditional techniques and systems do also occur in India, such as at Auroville. They are, however, still so rare that it may not yet be called a trend.

When it comes to the use of the oldest of all building materials, unfired clay, it is believed that one half of the world's population continues to live or work in earthen buildings.[18] For many of these people who live in hot and dry climates, unfired clay is not only the least costly material, but also the most practical because it insulates against the large diurnal temperature shifts far better than reinforced concrete. In addition, unfired clay has other benefits. Little energy is consumed in its quarrying and preparation as a building material, and it can be recycled by simply letting it wash back onto the ground. Earthen construction has even proven to be suitable in the wet British and Northern European environment, where centuries-old English "cob" houses continue to be lived in. In the eastern part of Germany after the Second World War, when other materials were simply unavailable, blocks of socialist housing were constructed of pisé. They continue to be occupied and are now recognized as historically significant.

Kashmir's traditional construction utilizes a combination of unfired and fired clay, stone and timber. In urban areas, new construction in *dhajji dewari* has been gradually dying out, and the timber lacing found in *taq* is a nearly forgotten art. In rural areas, these construction traditions, including variations of *taq* and *dhajji* along with timber log and sawn board structures, are still very much alive, but reinforced concrete is now embraced by increasing numbers of people.

The 2005 earthquake, of course, has put all different types of construction to the test, especially in the damage districts on both sides of the Line of Control in Kashmir. On the Indian side, examples of *dhajji* construction, as reported by reconnaissance teams, did show good performance, especially when compared with unreinforced rubble stone masonry construction, but on the Pakistan side, where the earthquake damage and loss of life was considerably greater, the reconnaissance teams remained focused on the devastated urban areas, failing to note that the *dhajji* houses in rural areas performed well –

5.5c Owner-builder in the village of Thub, Pakistan Administered Kashmir, constructing a new house of dhajji construction to replace one of rubble stone destroyed in the 2005 earthquake. See also Figures 3.4d & 3.4e.

5.5d One of the local workers infilling the timber frame with rubble stone and mud mortar, followed with mud plaster, in the new dhajji house in Topi visible in Figures 4.2.4c and 5.5f & g, replacing a destroyed rubble stone house.

most likely because the traditional technique had gone out of use in recent years in favor of unreinforced rubble stone, which performed very badly in the earthquake.

Subsequent to the earthquake on both sides of the border, increasing numbers of people who lost their rubble stone houses, rebuilt them using *dhajji* construction. For example, in two villages on the Pakistan side documented by architect Maggie Stephenson of UN-HABITAT and engineer Kubilây Hiçyılmaz, during their technical assistance work, the villagers reconstructed their houses using the local form of *dhajji* construction (Figures 5.5f and 5.5g) after observing that the few *dhajji* houses in the village survived (Figure 5.5e and also see Figure 3.4d). Their collapsed houses had all been of rubble stone. The one reinforced concrete building that had been constructed in one of the villages, a three-storey commercial building, also collapsed (see Figure 3.4c). Today, over 80% of the reconstructed houses in both of these villages are of *dhajji* construction.[19]

What is most significant about this phenomenon is the fact that the people began rebuilding with *dhajji* construction before the Pakistani government agreed to provide them with any financial assistance. Based on the reports of their engineering consultants, and on the requirements made by the World Bank, the government had said that only houses that complied with the government's engineering requirements for reinforced masonry or concrete construction would be funded. Thus, the villagers undertook reconstruction with *dhajji* with little hope of government disaster assistance. They did this for two reasons. First, their remote site made access to steel and cement difficult despite government assistance. Perhaps even more important, though, was their own observations that the only houses to survive the earthquake were of *dhajji* construction, the one concrete building having collapsed.

Eventually, as described in Sections 3.3 and 3.5, in November of 2006, over 13 months after the earthquake, the government agreed to allow *dhajji* construction for reconstructed houses. The report by the Earthquake Reconstruction and Rehabilitation Authority (ERRA) on this adoption, quoted in Section 3.5, shows that, although their primary motivation was the lack of concrete and steel in rural areas, the government also was persuaded by their increased understanding of the earthquake resistance of the *dhajji* system. Maggie Stephenson of UN-Habitat reports that "as of January 2009, at least 100,000 houses have been constructed in Kashmir and NWFP using *dhajji* construction. This represents a remarkable revival of the local construction technique. Hopefully initiatives to sustain and regenerate the forest long term will ensure *dhajji* remains a viable option for communities in future."

As described in Section 3.3.2, in rural areas of the Northern Areas and the Northwest Frontier Province *bhatar* construction is preferred. One rather compelling (but unfortunate) reason in addition to earthquake safety is that, as the local people have said, "it stops bullets better than *dhajji*". In 2007, despite opposition by the World Bank, the Pakistan Government approved new *bhatar* construction as compliant for earthquake reconstruction funding. ERRA then co-published, with UN-HABITAT and SDC, *Bhatar Construction, An Illustrated Guide for Craftsmen*, prepared by Tom Schacher of the SDC (see Section 3.3.2). This guide is now used to train owner-builders and craftsmen. (A copy is available at: www.traditional-is-modern.net.)

New construction is where the concept that "traditional is modern" is put to the test. A living society with a vibrant architectural culture must incorporate changes over time, but at the present moment, the ongoing changes in Srinagar and the other communities most often represent a rejection of the past and its replacement with imported architectural language and forms of construction, so a rediscovery of the technical value of the local vernacular can help to reestablish a sense of regional identity. The question of whether premodern construction technologies are feasible for new construction is very different from proposing that heritage structures can be preserved and restored, yet both initiatives go together. By reviving a respect for traditional construction on a technical level (particularly for protection against earthquakes), a renewed respect for its cultural value can be reclaimed as well. When that happens, the craft of construction can take its rightful place alongside the other indigenous arts and crafts that have long distinguished Kashmir — many of which, such as carpet weaving and woodworking, are related to making buildings into works of art.

NEW *DHAJJI* BUILDINGS IN PAKISTAN ADMINISTERED KASHMIR

5.5e *This older house in Topi, near Bagh, of* dhajji *construction survived the earthquake with damage only to an unreinforced stone wall in the back that lacked timber lacing. It became the inspiration for the construction of new* dhajji *houses including the one shown below.*

5.5.f & g *This house in the village of Topi was photographed in 2006 while being reconstructed after the 2005 earthquake that destroyed the rubble stone house seen on the right in 5.5g below. In 5.5g, the owner of the house is standing to the right of his chief carpenter. Image 5.5f was taken in June 2006 by Maggie Stephenson, UN-HABITAT, and the image 5.5g, which shows the wall after the nogging with stone in mud mortar was complete, was taken by the author in October of the same year.*

5.5h *A new house with* cator *and cribbage timber lacing. The Balti word* cator *has the same meaning as the Pashto word* bhatar. *The regional difference is to have the extra reinforcement provided by the cribbage in the corners. This house was constructed for the former owner of the historic Altit Fort by the Aga Khan Cultural Services Division. Photograph by Tom Schacher, Swiss Agency for Development and Cooperation (SDC).*

5.5i *This new barn of* bhatar *with dry-laid stone masonry construction is in a mountain village near Battagram (in Gari Nwab Said) in the Northwest Frontier Province of Pakistan. The site is remote from the road, one hour by car, then another half-hour by foot away from Battagram, making reinforced concrete construction impossible. Photograph taken in 2007 by Tom Schacher, Swiss Agency for Development and Cooperation (SDC).*

5.6 Sustainability and Green Architecture

To demolish traditionally built houses, which could easily be refurbished, is environmentally damaging. The embodied energy in each of such houses is equivalent (according to the Building Research Establishment) to 4000 gallons of petrol—enough to send a Ford Mondeo round the earth five times. Those fossil fuels have already been burnt and the CO_2 is already in the atmosphere. So why repeat the process?

Quinlan Terry, Architect, UK[20]

"Sustainability" has become a contemporary mantra. All building construction utilizes resources, but traditional construction techniques emerged in times when resources were much more localized, and the conservation of such resources was thus embedded in the culture. Exotic materials such as marble, ceramic tile, and bronze could only be used for monumental buildings like maharajas' palaces and the Taj Mahal, not the common buildings that make up the city. Kashmiri traditional construction uses materials that are locally available and largely renewable, namely wood, stone, earth, lime and both fired and unfired clay. Now, after the arrival of reinforced concrete, people are quick to explain that "timber is too expensive", when in fact it is the most energy-efficient and sustainable structural material available today.

Sustainability in architecture has several aspects. On the one hand, materials used in construction are sustainable if they are renewable or abundant, are locally available or light and compact enough to make for economical transport, and use little fuel, with low emissions, during their manufacture into building materials. On the other hand, energy and resources used after a building is constructed and in use, such as for heat and light, must also be considered.

An example of sustainability in climatic building qualities is the Kashmiri *taq* buildings. They are made of earth, fired brick and timber, and traditionally finished inside with layers of mud and grass. These buildings have proven to be more energy efficient and comfortable than many of the modern buildings of brick and concrete, which in Kashmir are often uninsulated, despite the fact that energy for heat is expensive and often in short supply. The traditional houses depended on the insulation value of natural clay mortar and unfired clay masonry on the interior of the walls, but these can make a significant difference. Increasingly residents of concrete buildings have complained of bone and joint problems because of the effect of the hard cold floors and walls on their bodies during the winter. This problem is not unique to Kashmir.[21]

5.6a *This pre-industrial technique of sawing timbers into boards is still common in parts of Kashmir. Photograph by Abbid Hussain Khan, 2006, INTACH.*

5.6b *1979 view of the timber harvest in Kashmir. This is a rare sight today as the mature trees have been depleted without being replanted. Photo by Tom Dolan.*

5.6c This cement plant near Pampore, photographed in 2007, is shown spewing mountain-obscuring smoke into the atmosphere. It is only one of a total of eight cement plants around Srinagar.

5.6d Tar is shown here being burnt on the street next to Srinagar's Jama Masjid to heat it for road repairs, without regard for the toxic air pollution it creates. Photographed in May 2007.

With the increasing consciousness of the need for both energy conservation and a reduction in greenhouse gases that are causing global warming, a return to the appropriate and sustainable use of timber as a major building material needs to be seen as an important part of a modern approach to building construction. As observed by Prof. B. V. Venkatarama Reddy, of the Indian Institute of Science Bangalore: "Currently, the production of building materials contributes 22% of all CO_2 emissions because of the preponderance of the use of reinforced concrete and fired brick masonry. This amount could be reduced if increased amounts of timber were to be grown and used. If augmented with an increased use of unfired earth, particularly in rural areas instead of concrete and fired brick, the reduction will be larger. The earthquake-resistant properties of timber construction only reinforce the urgency of this shift."[22]

In the Vale of Kashmir, this situation is most acute. Within the last quarter of a century eight cement plants have sprung up, and the digging for limestone to supply the raw material for them is disfiguring mountains near Pampore, making them resemble girdled trees. As of 2007, one such illegal quarry threatens to wipe out one of the world's most important sites for geological research.[23] The massive pollution from these plants, together with the mushrooming automobile and truck traffic (including the military convoys that daily traverse the valley) increasingly obscures the iconic ring of mountains that surround the valley, and it progressively threatens the environment for the people who live there – as well as repelling the tourists who are an important source of local livelihood (Figures 5.6c and 5.6d).[24]

5.6.1 THE NEED FOR A CULTURE OF MAINTENANCE: In many communities around the world, it is the oldest houses that are now significantly more valuable than the newer ones. In the United States, for example, the 250-year-old wooden houses in New England are particularly prized (see Figure 5.5a, for examples). In a recent study in Britain, it was found that on average, a pre-1919 house is worth 20% more than an equivalent house of more recent vintage. The older the home is, the greater the added value (Rypkema, 2005).

By contrast, in parts of Kashmir, "An old building is a building of only 50 years", UN-HABITAT architect Maggie Stephenson observes. "There is almost no culture of maintenance and repair work, so it can be assumed that at any given time there is a high proportion of buildings in an advanced stage of deterioration. Replacement, and thus the consumption of timber, has been extravagant." She goes on to say that the repair work advocated in Chapter 4 "will require the development of a corps of people with sophisticated skills", and that "most artisans prefer easier and more profitable new work".

In cities like Srinagar and Baramulla, the situation may not be as sharply defined as in the rural areas she has come to be familiar with, but the point is an important one in the discussion of sustainability as well as conservation of cultural heritage. The two issues can go hand-in-hand. When buildings are conserved, the precious materials of which they are made do not have to be re-harvested, mined or manufactured, and trees planted can be left to mature for the future. Then when they are cut, there will be a sufficient amount of heartwood to provide for durable repairs. When repairs are long-lasting, one tree can suffice when many would be needed for a new structure.

However, there is more than materials involved in conservation work. Buildings can only survive for extended life-spans if they are properly maintained, and such repair work does require specialized craft skills. Kashmir has long been famous for its crafts, so skills relevant to the conservation of buildings are already embedded in the culture, but the "culture of maintenance" does need to be embraced and encouraged through education and social support.

Historically, masons and carpenters have enjoyed almost mythical levels of social status and respect, but over the course of recent decades, much of this has been eclipsed by changes to the construction industry and the downgrading of traditional skills. In recent years in many countries, historic preservation has helped to reestablish these skills and status.

In Japan, for example, not only are historic buildings designated as landmarks, but the Japanese also have conferred the title "Living National Treasure" upon some 70 of their master artists and craftsmen, who are thus charged with passing on the country's artistic heritage to future generations. In Kashmir, a land already noted for its indigenous crafts, the skills involved in the creation and repair of traditional buildings can provide an unparalleled opportunity to expand the skilled job base and lift more people out of poverty – and be a source of honour.

5.6e Hōryūji (Hōryū Temple) near Nara, Japan is the site of the world's oldest wooden buildings, the 6th-century Kondō and 5-storey pagoda. This is a view of the magnificent timber facade of the Daikōdō, dating from the 8th century. In Japan, the maintenance of these and other timber relics has been carried to the level of a fine art, and in so doing, craft traditions have survived which enrich society outside the temples as well.

5.6f This view of a commercial street in Mingaora, Pakistan is typical of many in both modern India and Pakistan, with no more thought put into the town planning than has been put into seismic safety. The unregulated roadside commercial construction, absent even of sidewalks, is antithetical to traditional urban life and community.

The Mar Canal, Srinagar

5.7a The "Mar Qual" (Mar Canal) ca. 1864, by Samuel Bourne. Photo: British Library

5.7b This canal was filled and the buildings along it destroyed for a two-lane road in the 1970s, as seen here in 2005.

By the 19th century, the Mar Canal (*Nalla Mar*) had become a tourist attraction. In 1889, British traveler and writer Isabella Lucy Bird described it:

"Several hours were spent in a slow and tortuous progress through scenes of indescribable picturesqueness – a narrow waterway spanned by sharp-angled stone bridges, some of them with houses on the top, or by old brown wooden bridges festooned with vines, hemmed in by lofty stone embankments into which sculptured stones from ancient temples are wrought, on the top of which are houses of rich men, fancifully built, with windows of fretwork of wood, or gardens with kiosks, and lower embankments sustaining many-balconied dwellings, rich in colour and fantastic in design, their upper fronts projecting over the water and supported on piles…. and all the other sights of a crowded Srinagar waterway, the houses being characteristically distorted and out of repair." [25]

These comments were published only three years after the earthquake of 1886. Today, this canal through the heart of Srinagar is gone – replaced by a road that can never be the subject of such a poetic observation.

5.7 CONCLUSION

In the 17th century, Emperor Jahangir (reigned 1605-27) reportedly said of Kashmir: "If there is a heaven on earth, this is the place, this is the place, this is the place." Ever since, Kashmir has beguiled visitors and residents alike. Srinagar, with its navigable river and the network of canals threading through the dense city fabric, prompted the British writer Isabella Lucy Bird to write in 1889: "Never had this Venice of the Himalayas, with a broad rushing river for its high street and winding canals for its back streets, looked so entrancingly beautiful."

Today, these superlatives would inevitably be tempered by the environmental impact in recent decades of unregulated redevelopment and the filling of the canals. In the 1970s, the fabled Mar Canal (Figures 5.7a and 5.7e) was filled and replaced with a two-lane road lined with concrete shops (Figure 5.7b). There is just one canal left that preserves a piece of what had been a famous urban landscape. The surviving canal extends from Dal to Nagin Lakes through Rainawari (Figure 5.7f). This is now severely polluted, and many of the houses that line it are abandoned, but its visual magic is still there. It can be preserved, but doing so will require action by citizens and leaders willing to rediscover Kashmir's remarkable heritage of vernacular architecture and construction and embrace this last opportunity to pass it on to future generations.

Ironically, the recent years of conflict have not restrained redevelopment in the historic centre of Srinagar as one might have expected. Instead, the effect has been the opposite. There are currently few land use controls and little oversight, so change continues to be rapid, but there is a lack of collective vision and planning to control its impact.

It is relevant to take note of how modern development has affected Kathmandu, Nepal, which shares Srinagar's stature as an historic Himalayan capital city. With seven UNESCO World Heritage sites within the Kathmandu Valley, Kathmandu is more widely recognized by the world community for its architectural heritage. However, the listing of these major sites has not prevented the same kinds of problems from rampant unregulated development, as can be seen in Appendix 1. This danger from unregulated and unrestrained "modernization" was addressed by international economic development specialist Donovan Rypkema in a 2005 keynote address to Europa Nostra, when he observed: "While economic globalization has many positive effects, cultural globalization has few if any benefits, but has significant adverse social and political consequences in the short term and negative economic consequences in the long term. If cities are to succeed in economic globalization, they will have to be competitive worldwide. However, their success will be measured not just by their ability to foster economic globalization, but equally in their ability to mitigate cultural globalization. In both cases, a city's cultural heritage will play a central role. The modernization

5.7c View in October 2005 of canal leading from Dal Lake in Nowpora, Srinagar. This view captures some of the magic of Srinagar's fabled canal and shallow lake environment.

5.7d The same view in April 2007. The clearing of all the trees, with little distinction made between those that had been there for generations and those that were recent intrusions into the lake, was done with little regard for the aesthetics and culture of the environment. The communications tower is equally insensitive. The Srinagar lakes are more than natural environments; they have been shaped and occupied by human settlements for centuries, and they constitute cultural landscapes of the highest significance.

5.7e The "Mar Qual" (Mar Canal) ca. 1864, by Samuel Bourne. Photo: British Library

5.7f The Rainawari Canal, although now badly polluted and with buildings in disrepair, still has some of the character of the lost Mar Canal and holds the greatest potential as a future tourist attraction. It is the one place where the historic vernacular houses and traditional construction come together with the water-infused landscape that makes Srinagar unique. Saving this is of critical importance — the last opportunity to preserve an icon of the heritage of Srinagar, a city once known as "the Venice of the Himalayas". [26]

of cities in terms of infrastructure, public health, convenience, and quality of life does not necessitate the 'Americanization' of the built environment. An imitative strategy for the built form quickly leads a city from being 'someplace' to 'anyplace,' and the distance from 'anyplace' to 'no place' is short indeed." (Rypkema 2005).

On the other hand, there are some encouraging developments. The Indian National Trust for Art and Cultural Heritage (INTACH) has completed an exhaustive reconnaissance and published a multi-volume comprehensive survey of heritage structures. The organization has also progressed with further planning and preservation initiatives. In 2007 Srinagar was placed on the "Watch List" of the World Monuments Fund, an international NGO based in New York City. The Watch List was established to attract attention and concern to places of world significance under threat. Finding a way to build on these initiatives will be the challenge of the coming years, but these steps are essential in beginning to bring the issues to public attention and creating awareness not only that something should be done, but that it can be done and that people's lives will be the better for it.

Srinagar's and Kashmir's vernacular architecture with its earthquake-resistant construction is of central importance to all of these issues of cultural preservation, sustainable economic growth and quality of life. One only needs to look at the image on the Khyber Cement Company billboard in the centre of Srinagar (Figure 5.3d and 5.7h), presenting the future of Srinagar as a version of New York or Hong Kong, to understand why. For tourists and residents alike, this is not an image related to the culture and people of Kashmir.

The earthquake performance issue is in fact fundamental to the point that "traditional is modern". These are not just old buildings waiting to be scrapped and replaced, with a few worth setting aside in a theme park or museum: they are buildings that embody distinctly modern construction features – features that can save lives once they are fully researched, understood and embraced. These buildings are also significantly more sustainable than modern construction based on steel, concrete block and reinforced concrete. If old buildings built by hand with few tools, little formal education, and even less money can outperform new buildings of modern materials and technology in response to one of the largest forces that nature can throw at them, then indeed there is something to learn from them, and from the people and culture that brought them into being. Such knowledge can then save lives in the future. Thus, traditional *is* modern.[27]

When people understand that traditional pre-industrial materials and methods of construction can be embraced as a source of ideas on how to make new buildings better, they may rediscover parts of their heritage that increasingly have been spurned as backward. Construction technology is an unsung and little known area of architectural and cultural history, but for vernacular buildings, their unvarnished and unadorned construction forms a large part of their cultural significance.

5.7g *The architecture and construction of this new house being built in 2005 in the heart of a historic Srinagar urban neighbourhood shows the change that has taken place in Kashmir vernacular. The masonry laid in cement mortar lacks the texture of the Maharaji brick surfaces, the windows are small and lack any of the ornamental woodwork that grace the older houses, and the house fills its site without the timber brackets and other features of the older houses that give character to the area. The visual effect is oppressive, and diminishes the environmental quality of the surrounding heritage buildings.*

A City's Heritage Can Also Be Its Future

5.7h This is the image of a future reinforced concrete city that the Khyber Cement Company imagines for "modern" Srinagar. This cityscape adorns the company's billboards, which are located at strategic intersections throughout Srinagar, as seen in Figure 5.3d. The image seems to be a combination of Hong Kong combined with New York City. One thing it definitively is not, is Srinagar. The more Srinagar is taken in this direction, the greater will be the loss to its cultural and aesthetic uniqueness and value, which, as Donovan Rypkema has pointed out in Section 5.7, will have severe economic as well as social consequences.

5.7i Restored Fachwerk timber and masonry buildings in Weilburg, Germany. These buildings, and many others like them throughout Europe have been restored to continue to be lived in and used.

5.7j New timber frame of douglas fir being constructed in 2007 by master carpenter Klaus Becker in Greifenstein Ulm, near Frankfurt, Germany for a new warehouse building of Fachwerk construction.

5.7k Students at the Shah-e-Hamdan Memorial Trust School in Pampore, photographed on April 30, 2007. Perhaps the most important component of the heritage preservation effort is education: the education of the decision-makers today, but also – and perhaps more importantly – of the youngsters in school who, with each passing year, will otherwise grow more distant from the unique culture of Kashmir.

5.7l (below) Dal Lake, looking west towards Srinagar, April 2007.

The buildings described here are seemingly unpretentious, weak, insubstantial buildings that have been renewed for generations, but they now qualify as monuments. In the course of history, many masonry buildings have collapsed on their occupants during earthquakes; but it is essential to examine those which have survived, even when today's conventional wisdom predicts that they would not.

Many alive today have heard the refrain, "You cannot stop progress," with its assumption that "progress" represents an inevitable movement toward a better life. Lately, with the destruction caused by the 2004 Sumatra earthquake and tsunami, Hurricane Katrina in 2005 and the 2008 Wenchuan earthquake in China, this view has ceased to be so compelling. The repeated collapse of thousands of reinforced concrete schools, homes, and apartment houses in earthquakes around the world continues to provide further evidence of the fallacy of a belief in eternal progress.

Civilization rests on humble as well as grand contributions. The *hımış* and *dhajji dewari* structures that have been found standing amidst the earthquake ruins of the modern buildings around them do not mock their modern neighbours laid low, but rather, they quietly encourage us to shed some of the arrogance and over-confidence that brought the now collapsed buildings into being, forcing us to re-examine the roots of our civilization for ideas of how to build better in the present, even while we explore new and more modern materials and forms for the future.

> *Like some supremely beautiful woman, whose beauty is almost impersonal and above human desire, such was Kashmir in all its feminine beauty of rivers and valleys and lakes and graceful trees. And then another aspect of its magical beauty would come into view, a masculine one, of hard mountains and precipices, and snow capped peaks and glaciers, and cruel and fierce torrents rushing down to the valley below. It had a hundred faces and innumerable aspects, ever changing, sometimes sad and full of sorrow…it was like the face of the beloved that one sees in a dream and that fades away on awakening…*

> Jawaharlal Nehru, 1936[28]

Let us preserve and rebuild these qualities for present and future generations in Kashmir.

Access to Timber as a Building Material Requires the Conservation of Forests

- INDIA: The rapid rate of deforestation has slowed and in recent decades has not increased overall, as reported by the UN Food and Agriculture Organization; but the area has already largely been depleted of its timber.[30]

- PAKISTAN: The forests, which as of 2006 cover only 4.8% of the country, are disappearing at the rate of 4% to 6% a year, one of the highest rates in the world. Most of the wood is cut illegally and consumed for fuel, and the high price of usable lumber also encourages illegal cutting. It is predicted that Pakistan's woody biomass may be totally consumed within the next 10–15 years.[29]

Fuel Usage in Production of Construction Materials

The manufacture of building materials produces 22% of all CO_2 emissions. The energy consumed to produce them is equivalent to 150 million tons of coal. The production of cement (a major source of pollution and energy cost) has risen from 20 million metric tons annually in 1981 to 110 million in 2003, with a capacity in 2003 of 140 million. Burnt brick consumes topsoil at the rate of 1,000 sq km/year (Venkatarama Reddy, nd).

While some of the materials used in traditional construction require fuel use and carbon dioxide generation for their manufacture and use, reinforced concrete requires much more fuel by comparison and results in significantly more carbon dioxide emission, which contributes to global warming. The greater amount of fuel is needed because of the significantly higher temperatures required than for lime mortar or clay brick, as well as the need for large volumes of concrete and steel for columns, beams and floor slabs. In addition, the mortar used for conventional infill walls requires even more cement, and the manufacture of cement itself releases carbon dioxide into the atmosphere from more than just the fuel to fire it because of the chemical reaction resulting from the conversion of calcium carbonate into calcium oxide.

- CONCRETE: Concrete is made from a mixture of limestone and silicates from shells or chalk, and shale, clay, sand, or iron ore fired at 1430°C to 1650°C (2600°F to 3000°F). The chemical reaction from the making of cement from limestone also releases carbon dioxide into the atmosphere.

- STEEL: The manufacture of steel goes through several heating and melting stages with the highest temperatures between 1400°C and 1600°C.

- FIRED BRICKS: The temperature for making standard fired bricks from clay is 900°C to 1100°C.

- LIME MORTAR AND PLASTER: Building lime is manufactured by burning limestone, but at 954°C and 1066°C, a much lower temperature than for concrete.

- WOOD: No fuel is required in its processing except if kiln-dried. The only fuel required is for cutting and transporting the harvested material, and for the manufacture of derivative products such as plywood and oriented strand-board.

- STONE: Fuel is only needed for quarrying and transport, and for the making of hydrated lime and/or cement for mortar.

- UNFIRED CLAY (ADOBE AND PISÉ): Fuel is only needed for digging and local transport. (Large-scale indiscriminate quarrying of unfired and fired clay for building material can result in the significant loss of agriculturally rich topsoil, but the best earth for earthen construction is generally not very good for agriculture.)

5.7m Baramulla resident Mr Zargar, 90 years old, seen standing in the window of his ca. 250-year-old house. He was thrown by the earthquake out of this window and broke his leg. He returned from the hospital to live in the earthquake-damaged house seen in Figure 4.3.1g where he was photographed in April, 2007.

APPENDIX 1

THE KATHMANDU, NEPAL COMPARISON

A.1&2 The restoration work at the 55 Windows Palace in Bhaktapur, Nepal is one of the many examples of the continuation there of a handcraft tradition.

To understand how changes in construction have affected Kashmir, it is worth comparing their impact on Nepal. Over the course of the last half-century, the capital city, Kathmandu, has undergone a transformation, even though the historic value of large parts of it has been recognized by being placed on UNESCO's World Heritage List. Two photographs (Figure A6 and A7) show this change profoundly.

Large amounts of money have flowed into the country from the developed world under the premise of improving the well-being of what has been viewed as an "underdeveloped" country, but much of this "development" has been destructive of both the environment and the culture of the place. As author Jürgen Schick Reports:

> *This "development" of Nepal according to Western standards…has occasioned within only forty years the almost total destruction of its landscapes, its way of life, its traditions and its cultural identity…Dollars flowed in by the millions. Nepal was overwhelmed at one blow with the blessings of the modern age: automobiles, paved roads, motorcycles, radios, telephones, computers, wristwatches, video equipment, cement factories, and even…television…The veil of magic that once lay over Nepal is now torn. In the Kathmandu Valley the signs of modernity are there for all to see. The snow peaks no longer glow with radiant clarity over the Valley. Now they shine but dimly through the grey-brown haze that the cement factory of Chobhar, a gift from the Federal Republic of Germany, spews out day and night from its unfiltered stack. Today the rivers and streams flowing through the Valley are no longer silver…Nowadays the Nepalese no longer build their beautiful ochre-red houses, let along their five-storeyed pagodas, out of brick, wood and packed earth, for now that cement is available, building is done in the "modern" style. Thus bunker-like structures of bare grey cement – hideously ugly monstrosities – are eating their way into green rice fields everywhere. Nowadays, finally, even the golden images of the gods no longer smile down from the temple walls [for] everywhere there are gaping holes where they have been pried loose and stolen.*

Jürgen Schick, *The Gods are Leaving the Country*, 1998

The fate of Kathmandu also serves to show that if the surviving heritage qualities in Srinagar, Baramulla and other heritage towns and cities in Kashmir are not recognized for their value now, they will likely disappear fast, and be replaced with a mediocrity of the kind of reinforced concrete buildings that can be found anywhere.

A.3 Compare the cultural richness seen in Figures A.1, 2 & 4 with this common scene of unregulated development of alien forms of concrete buildings in the fertile agricultural flood plane that surrounds Kathmandu.

A.4 Patan Durbar Square in Kathmandu – a World Heritage Site, February 2005.

A.5 Just a few blocks from the Patan Durbar Square World Heritage Site in Kathmandu, Nepal, the unregulated development of reinforced concrete buildings disfigures what had been an extensive heritage area.

A.6 & A.7 Two photographs of Kathmandu, Nepal. The top photo was taken in the late 1920s, and the bottom one in about 1990. The juxtaposition of these two images – both taken from the same location – dramatically shows what can happen to a city which began with a character that is very similar to that of Srinagar – and which today has been altered heavily by multi-storey buildings. Despite this, Kathmandu today is still well regarded for its heritage and architecture. But is it this kind of change we want to happen in Srinagar and the other cities of Kashmir? Source: Proksch, 1995.

Appendix 2

Glossary of Technical Terms

Action: Forces or moments that cause displacements and deformations.

Active fault: A fault for which there is an average historic slip rate of 1mm per year or more, and evidence of seismic activity within Holocene times (past 11,000 years).

Adobe: Spanish for "earth". This term has come to mean in English unfired clay masonry, where the clay is cast into blocks or bricks, as opposed to walls constructed in place out of earth, as in "cob" or pisé (rammed earth).

Balloon frame: A framing system (USA & Canada, ca. 1850-1950) in which studs and corner posts extend from the sill to the plate and the upper-storey floor joists are carried on ledgers or girts let into the studs. Now largely replaced by "platform frame" method.

Bearing wall: A wall that supports gravity loads of at least 200 pounds per lineal foot from floors and/or roofs.

Bed joint: The horizontal layer of mortar on which a masonry unit is laid.

Boundary member: A member at the perimeter (edge or opening) of a shearwall or horizontal diaphragm that provides tensile and/or compressive strength.

Braced frame: A vertical lateral-force-resisting element consisting of vertical, horizontal, and diagonal components joined by concentric or eccentric connections.

Building occupancy: The purpose for which a building, or part thereof, is used, or intended to be used, designated in accordance with the applicable building code.

Building performance level: A limiting damage state for a building, considering structural and nonstructural components, used in the definition of rehabilitation objectives.

Building type: (As used in earthquake engineering) A building classification that groups buildings with common lateral-force-resisting systems and performance characteristics in past earthquakes.

Capacity: The permissible strength or deformation for a component action.

Cator: The term used for a timber band in masonry construction in "*cator* and cribbage" construction in Northern Pakistan (Hughes, 2000).

Cob: A building material consisting of clay, sand, straw, water, and earth, that is built up in place, rather than cast into blocks (adobe) and it is not rammed after placement, as with "rammed earth" (pisé).

Collar joint: Vertical longitudinal joint between wythes of masonry or between masonry wythe and backup construction; can be filled with mortar or grout.

Collector: A member that transfers lateral forces from the diaphragm of the structure to vertical elements of the lateral-force-resisting system.

Component, flexible: A component, including attachments, having a fundamental period greater than 0.06 seconds.

Component, rigid: A component, including attachments, having a fundamental period less than or equal to 0.06 seconds.

Components: The basic structural members that constitute a building, including beams, columns, slabs, braces, walls, piers, coupling beams, and connections; designated as primary or secondary.

Concentric braced frame: Braced frame element in which component worklines (load-resisting members) intersect at a single point or at multiple points such that the distance between intersecting components, or eccentricity, is less than or equal to the width of the smallest member connected at the joint.

Confined masonry: A system of reinforced concrete frame construction where the infill masonry elements are constructed first and the concrete components are then cast in place around them. The formwork necessary for the concrete components is thus less. This system is most often used for low-rise structures, but in some areas, including Mexico, multi-storey mid-rise buildings are being constructed with this method. It has generally demonstrated good resistance to earthquakes as long as the infill panels are not too large. (*Image, credit: Tom Schacher*)

Connection: A link that transmits actions from one component or element to another component or element, categorized by type of action (moment, shear, or axial).

Connectors: Nails, screws, lags, bolts, split rings, and shear plates used to link wood components to other wood or metal components.

Coupling beam: A component that ties or couples adjacent shearwalls acting in the same plane.

Creep: Slow, more or less continuous movement occurring on some faults. Creep does not cause shaking.

Cribbage: A system of timber construction where heavy timbers are placed crosswise to each other to form a box, which in Kashmiri as well as northern Pakistani construction is then filled with masonry.

Crosswall: An interior partition that extends from the floor to the underside of the floor above, securely fastened to each and capable of resisting lateral forces based on its relative stiffness, strength, deformation and energy dissipation.

Damping: Any effect, either deliberately engendered or inherent to a system, that tends to reduce the amplitude of oscillations of an oscillatory system (see "viscous damping" and "hysteretic damping").

Dead load: The weight of the building materials that make up a building including its structure, enclosure and architectural finishes that are supported by the structure.

Decay: Decomposition of wood caused by action of wood-destroying fungi. The term "dry rot" is used interchangeably with decay.

Deformation: In engineering mechanics, deformation is a change in shape due to an applied force. This can be a result of tensile (pulling) forces, compressive (pushing) forces, shear, bending or torsion (twisting). Deformation is often described in terms of strain.

Demand: The amount of force or deformation imposed on an element or component.

Design-level earthquake: The earthquake which can reasonably be expected to occur at least once during the design life of the structure. Typically, the US Standards accept it as an earthquake with probability of 10% in 50 years' life of a structure.

Diagonal bracing: Inclined components designed to carry axial load, enabling a structural frame to act as a truss to resist lateral forces.

Diaphragm: A horizontal structural element (usually the floor structure) used to distribute inertial lateral forces to vertical elements of the lateral-force-resisting system.

Diaphragm chord: A component provided to resist tension or compression at the edges of a diaphragm.

Ductility: The property of a material that allows it to sustain inelastic deformations without breaking. The ability of a structure to sustain its load-carrying capacity and dissipate energy when subjected to cyclic inelastic displacements during an earthquake.

Ductility factors: Constants used in engineering calculations that are included in building codes or structural engineering manuals that are assigned based on the ductility of the materials that make up the individual elements of a structural frame.

Ductile reinforced concrete: Reinforced concrete structural design with detailing to meet code provisions for ductile behaviour in earthquakes.

Earthquake: Ground shaking caused by a sudden movement on a fault or by volcanic disturbance.

Earthquake fault: Plane zone along which earth materials on opposite sides have moved differentially in response to tectonic forces. On the earth surface it is expressed as a line.

Eccentric braced frame: Braced frame element in which component worklines (load-resisting members) do not intersect at a single point and the distance between the intersecting components, or eccentricity, exceeds the width of the smallest member connecting at the joint.

Effective damping: The value of equivalent viscous damping corresponding to the energy dissipated by the building, or element thereof, during a cycle of response.

Effective stiffness: The value of the lateral force in the building, or an element thereof, divided by the corresponding lateral displacement.

Elastic: Capable of returning to its original length or shape after being stretched.

Elastic deformation: This type of deformation is reversible. Once the forces are no longer applied, the object returns to its original shape.

Elastic limit: The limit of being stretched before permanent (inelastic) deformations occur.

Element: An assembly of structural components that act together in resisting forces, including gravity frames, moment-resisting frames, braced frames, shearwalls, and diaphragms.

Energy dissipation: The loss of energy from a system due to the conversion of work into other forms of energy, such as heat due to friction in a structural system.

Epicentre: The epicentre is the point on the earth's surface that is directly above the hypocentre or focus, which is the point where an earthquake originates.

Flexible connection: A link between components that permits rotational and/or translational movement without degradation of performance, including universal joints, bellows expansion joints, and flexible metal hose.

Flexible diaphragm: A diaphragm where the maximum lateral deformation along its length is more than twice the average inter-storey drift.

Foreshock: An earthquake that precedes the largest quake ("mainshock") of an earthquake sequence. Foreshocks may occur seconds to weeks before the mainshock.

Frame action: The mechanism by which a moment-resisting frame resists lateral forces through the bending of the beams and columns between their connections.

Fundamental period: The highest natural period of the building in the direction under consideration.

Head joint: Vertical mortar joint placed between masonry units in the same leaf or wythe (layer of masonry).

Hysteretic damping (hysteretic behaviour): There is energy loss within a moving structure itself that is called hysteretic damping or structural damping. In hysteretic damping, some of the energy involved in the repetitive internal deformation and restoration to original shape is dissipated in the form of random vibrations of the crystal lattice in solids and random kinetic energy.

Hysteretic diagram: A hysteretic diagram (see below) represents the force and displacement of a building over a single cycle of ground motion causing a non-linear response in the building's superstructure (*Illustration source: FEMA 451*).

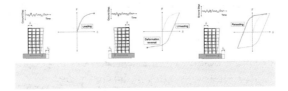

Immediate occupancy performance level: Building performance that includes very limited damage to both structural and nonstructural components during the design-level earthquake. The basic vertical and lateral-force-resisting systems retain nearly all of their pre-earthquake strength and stiffness. The level of risk for life-threatening injury as a result of damage is very low. Although some minor repairs may be necessary, the building is fully habitable after a design-level earthquake, and the needed repairs may be completed while the building is occupied.

Inelastic deformation (or plastic deformation): This type of deformation is not reversible. However, an object in the plastic deformation range will first have undergone elastic deformation, which is reversible, so the object will return partway to its original shape.

Infill: A panel of masonry placed within a steel or concrete frame. Panels separated from the surrounding frame by a gap are termed "isolated infills". Panels that are in full contact with a frame around its full perimeter are termed "shear infills".

In-plane wall: See shearwall.

Intensity: A measure of ground shaking describing the local severity of an earthquake in terms of its effects on the earth's surface and on humans and their structures. The Modified Mercalli Intensity (MMI) scale, which uses Roman numerals, is one way scientists measure intensity. A single earthquake would have many intensities.

Inter-storey drift: The relative horizontal displacement between two adjacent floors in a building; can also be expressed as a percentage of the storey height separating the two adjacent floors.

Khanqah: Sufi temple or shrine.

Lateral-force-resisting system: Those elements of the structure that provide earthquake resistance to a building through a combination of lateral strength, stiffness and energy dissipation capacity.

Life safety performance level: Building performance that includes significant damage to both structural and nonstructural components during a design-level earthquake, though at least some margin against either partial or total structural collapse remains. Injuries may occur, but the level of risk for life-threatening injury and entrapment is low.

Linear dynamic procedure (Modal response spectrum method): A response-spectrum-based modal structural engineering analysis procedure. It requires more engineering work, but with more precise results than the linear static procedure, which is appropriate for the analysis of buildings taller than 100 feet, buildings with vertical or geometric irregularities, and buildings where the distribution of the lateral forces departs from that assumed for the linear static procedure.

Linear static procedure (Equivalent static method): A lateral force structural engineering analysis procedure where the pseudo-lateral force is equal to the force required to impose the expected actual deformation of the structure in its yielded state when subjected to the design-level earthquake motions. It is appropriate for the analysis of simple, regular buildings.

Liquefaction: An earthquake-induced phenomenon in which saturated, loose, granular soils lose shear strength and liquefy as a result of increase in pore-water pressure during earthquake shaking.

Live load: The weight of all of the moveable contents of a building, including the occupants, furnishings, books and personal belongings that are supported by the structural system of the building.

Load path: A path through which seismic forces are delivered from the point at which inertial forces are generated in the structure to the foundation and, ultimately, the supporting soil.

Load sharing: The load redistribution mechanism among parallel components constrained to deflect together.

Low-strength masonry: Fired brick or stone masonry laid in weak mortars, such as cement-sand, lime-sand and clay-mud.

Magnitude: The measure of total energy released from an earthquake source, as determined from seismographic observations. The original earthquake magnitude scale was the Richter or "local" scale (ML), defined by Charles Richter in 1935, but it has limited range and applicability. Modern magnitude scales are based on the area of fault rupture times the amount of slip (seismic moment). The moment magnitude (MW) is the preferred magnitude scale, as it provides the most reliable estimate of the size of the largest quakes. For smaller quakes, ML and MW values are nearly the same. An increase of one unit of moment magnitude (for example, from 4.6 to 5.6) corresponds approximately to a 31.6-fold increase in energy released. By definition, a two-unit increase in magnitude—for example, from 4.7 to 6.7—represents an increase in energy released of 1,000 times. Quakes below magnitude 2.5 are not generally felt by humans. One earthquake can have only one magnitude.

Masonry: The assemblage of masonry units, mortar, and possibly grout and/or reinforcement; classified with respect to the type of masonry unit, including clay-unit masonry, concrete masonry, or hollow-clay tile masonry.

Maximum considered earthquake: It is the most severe earthquake that could happen in the design-life of a building Typically, the US Standards accept it as an earthquake with probability of 2% in 50 years' life of a structure

Mean return period: The average period of time, in years, between the expected occurrences of an earthquake of specified severity.

Moment frame: A building frame system in which members and joints are capable of resisting forces primarily by flexural capacity.

Non-ductile reinforced concrete: Reinforced concrete structure that is not detailed to meet the structural standards necessary for ductile behaviour.

Out-of-plane wall: A wall that resists lateral forces applied normal (perpendicular) to its plane. Walls generally are weaker in this plane.

Overturning: Action resulting when the moment produced at the base of vertical lateral-force-resisting elements is larger than the resistance provided by the building weight and foundation resistance to uplift.

Pancake collapse: The complete collapse of a building such that the floors are stacked like a pile of pancakes (*chapatis*).

Period of vibration: Time taken for the oscillator to undertake one complete vibration, returning to its original position and velocity.

Pisé: French term for rammed earth construction that is now commonly used throughout Europe and many other parts of the world.

Plate tectonics: The scientific theory that the earth's outer shell is composed of several large, thin, relatively strong "plates" that move relative to one another. Movements on the faults that define plate boundaries produce most large earthquakes.

Platform frame: Method of wood frame construction in which walls are erected on a previously constructed floor decks sequentially, floor-by-floor. *Dhajji* construction is usually a version of platform framing.

Plywood: A structural panel composed of plies of wood veneer arranged in cross-aligned layers bonded with adhesive cured upon application of heat and pressure.

Primary component: A part of the lateral-force-resisting system capable of resisting seismic forces.

Pseudo-lateral force: The calculated lateral force used for linear static procedure. The pseudo-lateral force represents the force required, in a linear analysis, to impose the expected actual deformation of the structure in its yielded state when subjected to design-level earthquake motions. It does not represent an actual lateral force that the building must resist in traditional code design.

Region of seismicity: An area with similar expected earthquake hazard. All regions are categorized as low, moderate, or high, based on mapped acceleration values and site amplification factors.

Retrofit (or "upgrade"): Strengthening an existing structure to improve its resistance to the effects of earthquakes.

Rigid diaphragm: A diaphragm that is sufficiently rigid that the maximum horizontal deformation along its length is less than half the average inter-storey drift.

Rupture zone: The area of the earth through which fault movement occurred during an earthquake. For large quakes, the section of the fault that ruptured may be several hundred kilometres in length. Ruptures may or may not extend to the ground surface.

Scarf joint: A joint in which two structural members are joined with long end laps and secured with keys or hardware to resist tension or compression (*see photos below by Kubilây Hiçyılmaz,*).

Secondary component: An element that is capable of resisting gravity loads, but is not part of the main energy-dissipating structure.

Seismic: Pertaining to, of the nature of, or caused by an earthquake.

Seismic demand: Seismic hazard expressed in the form of a ground-shaking response spectrum or accelogram with or without an estimate of permanent ground deformation.

Seismic evaluation: An approved process or methodology of evaluating deficiencies in a building which prevent the building from achieving a selected rehabilitation objective.

Seismic hazard: The potential for damaging effects caused by earthquakes. The level of hazard depends on the magnitude of likely quakes, the distance from the fault that could cause quakes, and the type of ground materials at a site.

Seismic risk: The chance of injury, damage, or loss resulting from seismic hazards. There is no risk, even in a region of high seismic hazard, if there are no people or property that could be injured or damaged by a quake.

Seismology: The science or study of earthquakes and their phenomena.

Shearwall: A structural component that resists lateral forces applied parallel with its plane, primarily by shear action. Also known as an in-plane wall.

Sheathing: Lumber or panel products that are attached to parallel framing members, typically forming wall, floor, ceiling, or roof surfaces.

Short column (captive column): Columns with height-to-depth ratios less than 75% of the nominal height-to-depth ratios of the typical columns at that level.

Site class: Groups of soil conditions that affect the site seismicity in a common manner. The soil types used are designated as A, B, C, D, E, or F.

Soft storey: A building storey that has significantly less stiffness than the storey above. Some buildings with parking at ground level (and thus fewer walls or columns) or an otherwise open ground storey have this condition. The term is sometimes also applied to a storey that has less strength than the one above, a condition that is more precisely termed a "weak storey".

Special procedure: Analysis procedure developed in the United States for unreinforced masonry (URM) bearing wall buildings with flexible diaphragms that properly characterizes the diaphragm motion, strength and damping, now accepted into the building code.

Strength: The maximum axial force, shear force, or moment that can be resisted by a component.

Stress and Strain: Stress: force that produces strain on a physical body. The intensity of stress is expressed in units of force divided by units of area. Strain: deformation of a physical body under the action of applied forces. (*Stress-strain diagram Source: Wikipedia*)

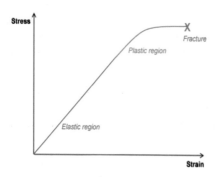

Strike-slip fault: A generally vertical fault along which the two sides move horizontally past each other. The most famous example is California's San Andreas Fault.

Structural components: Components of a building that provide gravity and lateral load resistance as part of a continuous load path to the foundation.

Structural system: An assemblage of load-carrying components that are joined together to provide regular interaction or interdependence.

Strut: A structural component designed to resist longitudinal compression.

Stucco: Exterior plaster (a term in use most commonly in North America).

Stud: Vertical framing member in interior or exterior wall.

Unreinforced masonry (URM) wall: A masonry wall containing no reinforcement or less than the minimum amounts of reinforcement as defined for masonry walls. An unreinforced wall is assumed to resist gravity and lateral loads solely through resistance of the masonry materials.

Vertical irregularity: A sudden change of strength, stiffness, geometry, or mass in one storey with respect to adjacent storeys.

Viscous damping: The linear elastic behaviour from friction of materials moving within their elastic range.

Work: A manifestation of energy; the transfer of energy from one physical system to another expressed as the product of a force and the distance through which it moves a body in the direction of that force; "work equals force times distance".

Wythe: A continuous vertical section (leaf or layer) of a masonry wall, one masonry unit in thickness (a term in use most commonly in North America).

Sources for Glossary definitions: FEMA 310; FEMA 256; NZS1170.5, IS1893-2002; Webster's Dictionary; Wikipedia and other sources.